CLINICAL CONNECTIONS

Gerard J. Tortora

Bergen Community College

WILEY

John Wiley & Sons, Inc.

Cover Photo: © Jupiter Images/Comstock Images/Alamy

Executive Editor	Bonnie Roesch
Program Assistant	Lauren Morris
Marketing Manager	Clay Stone
Senior Production Editor	Anna Melhorn
Snior Designer	Madelyn Lesure
Photo Editor	Saran Wilkin

To order books or for customer service please, call 1-800-CALL WILEY (225-5945).

ISBN-13 978-0-470-086667

Printed in the United States of America

10 9 8 7 6 5 4 3 2 1

Table of Contents

Table of Contents

Chapter 1 | Introduction

Most students can better understand normal anatomy and physiology by examining situations in which diseases or disorders impair the normal functioning of the body. The chapters in this book will provide you with concise discussions of major diseases and disorders that illustrate departures from homeostasis. There is a chapter devoted to each of the body systems that you are studying, beginning with Chapter 3 on the Integumentary system. Chapter 2 includes clinically relevant information linked to chemistry, cells, and tissues–levels of organization within the human body that are foundational to the system level.

Each of these chapters is organized into four main sections. The first section, **Tests, Procedures, and Techniques**, explores methods used within a clinical setting to diagnose or enhance a practitioner's knowledge about a patient's health or disease state. Next is a section on **Disorders**. These discussions highlight specific disorders and diseases that affect the relevant body system. Not only will these sections help you make a connection between normal anatomy and physiology and pathology and pathophysiology, but they will also answer some of the questions you have about certain medical problems. A third section, **Additional Clinical Considerations**, features discussions on topics of clinical relevance, but that are not necessarily concerned with a specific disorder or disease. An example would be *Sun Damage, Sunscreens and Sunblocks*, or *Regeneration of Heart Cells*. The final section of each chapter is devoted to **Medical Terminology**. This section will help you build your medical vocabulary. Terms include both normal and pathological conditions.

Homeostatic Imbalances

Homeostasis, the state of relative stability of the body's internal environment, is a dynamic condition. All structures, from the cellular level to the systemic level, contribute in some way to keeping the internal environment of the body within normal limits. As long as all the body's controlled conditions remain within certain narrow limits, body cells function efficiently, negative feedback systems maintain homeostasis, and the body stays healthy. Should one or more components of the body lose their ability to contribute to homeostasis, however, the normal equilibrium among body processes may be disturbed. If the homeostatic imbalance is moderate, a disorder or disease may occur; if it is severe, death may result.

A **disorder** is any abnormality of structure and/or function. **Disease** is a more specific term for an illness characterized by a recognizable set of symptoms and signs in which body structures and functions are altered in characteristic ways. A person with a disease may experience **symptoms**, *subjective* changes in body functions that are not apparent to an observer. Examples of symptoms are headache, nausea, and anxiety. *Objective* changes that a clinician can observe and measure are called **signs**. Signs of disease can be either anatomical, such as swelling or a rash, or physiological, such as fever, high blood pressure, or paralysis. A *local disease* affects one part or a limited region of the body; a *systemic disease* affects either the entire body or several parts of it.

The science that deals with why, when, and where diseases occur and how they are transmitted among individuals in a community is known as **epidemiology** (ep'-i-de-me-OL-o-je: *epi* = upon; *-demi* = people). **Pharmacology** (far'-ma-KOL-o-je; pharmac = drug) is the science that deals with the effects and uses of drugs in the treatment of disease.

Diagnosis of Disease

Diagnosis (di' –ag-NO-sis; *dia* = through; *-gnosis* = knowledge) is the science and skill of distinguishing one disorder or disease from another. The patient's symptoms and signs, his or her medical history, a physical exam, and laboratory tests provide the basis for making a diagnosis. Taking a *medical history* consists of collecting information about events that might be related to a patient's illness. These include the chief complaint (primary reason for seeking medical attention), history of present illness, past medical problems, family medical problems, social history, and review of symptoms and signs. A *physical examination* is an orderly evaluation of the body and its functions.

This process will include measurement of vital signs (temperature, pulse, respiratory rate, and blood pressure), and sometimes laboratory tests. In addition, several **non-invasive diagnostic techniques**—inspection, palpation, auscultation, and percussion—commonly used by healthcare professionals to assess certain aspects of body structure and function. In **inspection**, the first diagnostic technique, the examiner observes the body for any changes that deviate from normal. In **palpation** (pal-PA¯-shun; *palpa-* = to touch) the examiner feels body surfaces with the hands. An example is palpating a blood vessel (called an artery) to find the pulse and measure the heart rate. In

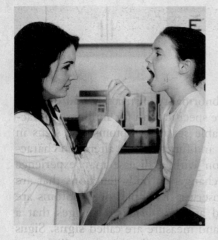

Inspection of oral (mouth) cavity

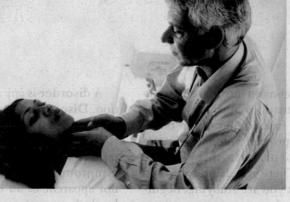

Palpation of lymph nodes in neck

Auscultation of lungs

Percussion of lungs

auscultation (aus-cul-TA¯-shun; *ausculta-* = to listen to) the examiner listens to body sounds to evaluate the functioning of certain organs, often using a stethoscope to amplify the sounds. An example is auscultation of the lungs during breathing to check for crackling sounds associated with abnormal fluid accumulation in the lungs. In **percussion** (pur-KUSH-un; *percus-* = to beat) the examiner taps on the body surface with the fingertips and listens to the resulting echo. For example, percussion may reveal the abnormal presence of fluid in the lungs or air in the intestines. It is also used to reveal the size, consistency, and position of an underlying structure.

Various kinds of **medical imaging** procedures allow visualization of structures inside our bodies and are increasingly helpful for precise diagnosis of a wide range of anatomical and physiological disorders. The grandparent of all medical imaging techniques is conventional radiography (x-rays), in medical use since the late 1940s. The newer imaging technologies not only contribute to diagnosis of disease, but they also are advancing our understanding of normal physiology. Table 1.1 describes some commonly used medical imaging techniques.

Death and Autopsy

Clinically, the loss of the heartbeat, absence of spontaneous breathing, and loss of brain functions indicate death in the human body. An **autopsy** (AW-top-se¯; *auto-* = self;-*opsy* = to see) is a postmortem (after-death) examination of the body and dissection of its internal organs to confirm or determine the cause of death. An autopsy can uncover the existence of diseases not detected during life, determine the extent of injuries, and explain how those injuries may have contributed to a person's death. It also may support the accuracy of diagnostic tests, establish the beneficial and adverse effects of drugs, reveal the effects of environmental influences on the body, provide more information about a disease, assist in the accumulation of statistical data, and educate healthcare students. Moreover, an autopsy can reveal conditions that may affect offspring or siblings (such as congenital heart defects). An autopsy may be legally required, such as in the course of a criminal investigation, or may be used to resolve disputes between beneficiaries and insurance companies about the cause of death.

TABLE 1.1

Common Medical Imaging Procedures

RADIOGRAPHY

Procedure: A single barrage of x rays passes through the body, producing an image of interior structures on x ray-sensitive film. The resulting two-dimensional image is a *radiograph* (RĀ-dē-ō-graf =), commonly called an *x ray.*

Comments: Radiographs are relatively inexpensive, quick, and simple to perform, and usually provide sufficient information for diagnosis. X rays do not easily pass through dense structures so bones appear white. Hollow structures, such as the lungs, appear black. Structures of intermediate density, such as skin, fat, and muscle,

appear as varying shades of gray. At low doses, x rays are useful for examining soft tissues such as the breast **(mammography)** and bone density **(bone densitometry).**

It is necessary to use a substance called a contrast medium to make hollow or fluid filled structures visible in radiographs. X rays make structures that contain contrast media appear white. The medium may be introduced by injection, orally, or rectally depending on the structure to be imaged. Contrast x rays are used to image blood vessels **(angiography)**, the urinary system **(intravenous urography)**, and the gastrointestinal tract **(barium contrast x ray).**

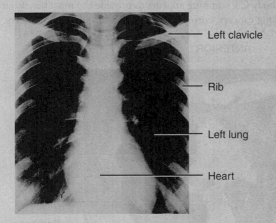

Left clavicle

Rib

Left lung

Heart

Radiograph of the thorax in anterior view

Mammogram of a female breast showing a cancerous tumor (white mass with uneven border)

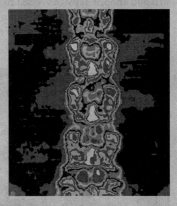

Bone densitometry scan of the lumbar spine in anterior view

Angiogram of an adult human heart showing a blockage in a coronary artery (arrow)

ntravenous urogram showing a kidney stone arrow) in the right ureter

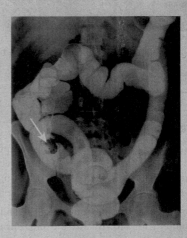

Barium contrast x-ray showing a cancer of the ascending colon (arrow)

⬤ **TABLE 1.1** CONTINUES ▶

TABLE 1.1 CONTINUED

Common Medical Imaging Procedures

MAGNETIC RESONANCE IMAGING (MRI)

Procedure: The body is exposed to a high-energy magnetic field, which causes protons (small positive particles within atoms, such as hydrogen) in body fluids and tissues to arrange themselves in relation to the field. Then a pulse of radio waves "reads" these ion patterns, and a color-coded image is assembled on a video monitor. The result is a two- or three-dimensional blueprint of cellular chemistry.

Comments: Relatively safe, but can't be used on patients with metal in their bodies. Shows fine details for soft tissues but not for bones. Most useful for differentiating between normal and abnormal tissues. Used to detect tumors and artery-clogging fatty plaques, reveal brain abnormalities, measure blood flow, and detect a variety of musculoskeletal, liver, and kidney disorders.

COMPUTED TOMOGRAPHY (CT) [formerly called computerized axial tomography (CAT) scanning]

Procedure: Computer-assisted radiography in which an x ray beam traces an arc at multiple angles around a section of the body. The resulting transverse section of the body, called a *CT scan*, is reproduced on a video monitor.

Comments: Visualizes soft tissues and organs with much more detail than conventional radiographs. Differing tissue densities show up as various shades of gray. Multiple scans can be assembled to build three-dimensional views of structures. In recent years, whole-body CT scanning has emerged. Typically, such scans actually target the torso. Whole-body CT scanning appears to provide the most benefit in screening for lung cancers, coronary artery disease, and kidney cancers.

ANTERIOR

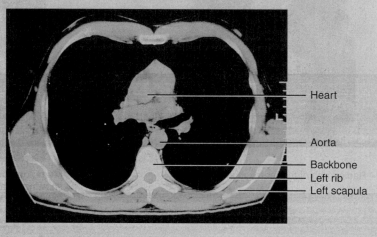

POSTERIOR

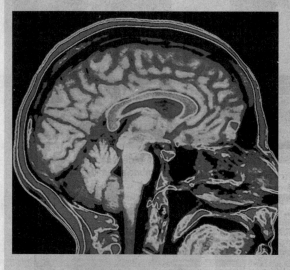

Magnetic resonance image of the brain in sagittal section

Computed tomography scan of the thorax in inferior view

ULTRASOUND SCANNING

Procedure: High-frequency sound waves produced by a handheld wand reflect off body tissues and are detected by the same instrument. The image, which may be still or moving, is called a *sonogram* (SON-ō-gram) and is reproduced on a video monitor.

Comments: Safe, noninvasive, painless, and uses no dyes. Most commonly used to visualize the fetus during pregnancy. Also used to observe the size, location, and actions of organs and blood flow through blood vessels **(doppler ultrasound).**

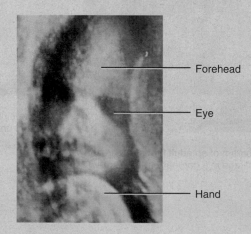

Sonogram of a fetus (Courtesy of Andrew Joseph Tortora and Damaris Soler)

POSITRON EMISSION TOMOGRAPHY (PET)

Procedure: A substance that emits positrons (positively charged particles) is injected into the body, where it is taken up by tissues. The collision of positrons with negatively charged electrons in body tissues produces gamma rays (similar to x rays) that are detected by gamma cameras positioned around the subject. A computer receives signals from the gamma cameras and constructs a *PET* scan image, displayed in color on a video monitor. The PET scan shows where the injected substance is being used in the body. In the PET scan image shown here, the black and blue colors indicate minimal activity, whereas the red, orange, yellow, and white colors indicate areas of increasingly greater activity.

Comments: Used to study the physiology of body structures, such as metabolism in the brain or heart.

ANTERIOR

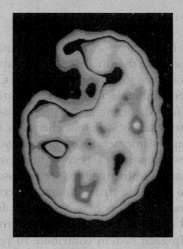

POSTERIOR

Positron emission tomography scan of a transverse section of the brain (darkened area at upper left indicates where a stroke has occurred)

RADIONUCLIDE SCANNING

Procedure: A *radionuclide* (radioactive substance) is introduced intravenously into the body and carried by the blood to the tissue to be imaged. Gamma rays emitted by the radionuclide are detected by a gamma camera outside the subject and fed into a computer. The computer constructs a *radionuclide image* and displays it in color on a video monitor. Areas of intense color take up a lot of the radionuclide and represent high tissue activity; areas of less intense color take up smaller amounts of the radionuclide and represent low tissue activity. **Single-photo-emission computerized tomography (SPECT) scanning** is a specialized type of radionuclide scanning that is especially useful for studying the brain, heart, lungs, and liver.

Comments: Used to study activity of a tissue or organ, such as the heart, thyroid gland, and kidneys.

Radionuclide (nuclear) scan of a normal human heart

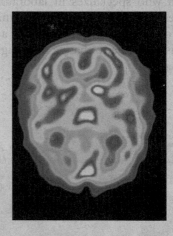

Single-photon-emission computerized tomography (SPECT) scan of a transverse section of the brain (green area at lower left indicates a migraine attack)

ENDOSCOPY

Procedure: The visual examination of the inside of body organs or cavities using a lighted instrument with lenses called an *endoscope.* The image is viewed through an eyepiece on the endoscope or projected onto a monitor.

Comments: Examples of endoscopy include colonoscopy, laparoscopy, and arthroscopy. *Colonoscopy* is used to examine the interior of the colon, which is part of the large intestine. *Laparoscopy* is used to examine the organs within the abdominopelvic cavity. *Arthroscopy* is used to examine the interior of a joint, usually the knee.

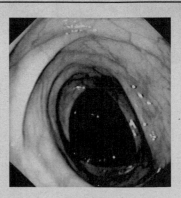

Interior view of the colon as shown by colonoscopy

Chapter 2 | Chemistry, Cells, and Tissues

Chemistry is the science of the structure and interactions of matter. Atoms are the smallest unit of matter and molecules are two or more atoms joined together. Certain atoms, (such as oxygen) and molecules (like DNA) are essential for maintaining life. Maintaining the proper assortment and quantity of thousands of different chemicals in your body, and monitoring the interactions of these chemicals with one another, are two important aspects of homeostasis.

Cells, the smallest living units, are enclosed by a membrane and are the basic structural and functional units of the human body. Humans have about 200 different types of specialized cells, which carry out a multitude of functions that help each system contribute to homeostasis of the entire body.

Tissues are groups of cells and the materials surrounding them that work together to perform a particular function, such as protection, support, communication with other cells, and resistance to disease. A *pathologist* is a physician who specializes in laboratory studies of cells and tissues to help other physicians make accurate diagnoses. One of the principal functions of a pathologist is to examine tissues for any change that might indicate disease.

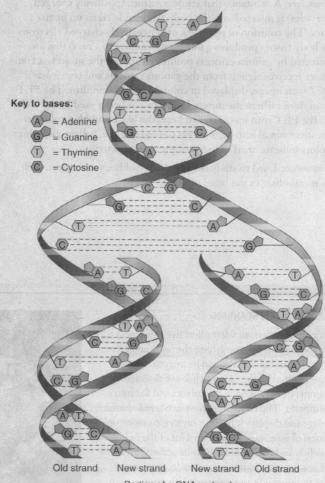

Key to bases:

(A) = Adenine
(G) = Guanine
(T) = Thymine
(C) = Cytosine

Old strand   New strand   New strand   Old strand

Portion of a DNA molecule

Typical cell structure

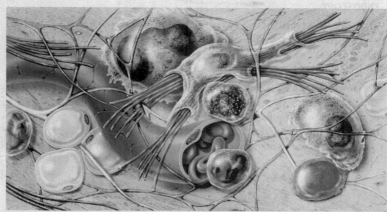

Representative cells & fibers in connective Tissue

6

Tests, Procedures, and Techniques

◼ DNA Fingerprinting

A technique called **DNA fingerprinting** is used in research and in courts of law to ascertain whether a person's DNA matches the DNA obtained from samples or pieces of legal evidence such as blood stains or hairs. In each person, certain DNA segments contain base sequences that are repeated several times. Both the number of repeat copies in one region and the number of regions subject to repeat are different from one person to another. DNA fingerprinting can be done with minute quantities of DNA—for example, from a single strand of hair, a drop of semen, or a spot of blood. It also can be used to identify a crime victim or a child's biological parents and even to determine whether two people have a common ancestor.

◼ Liposuction

A surgical procedure called **liposuction** (*lip-* = fat) or **suction lipectomy** (*-ectomy* = to cut out) involves suctioning small

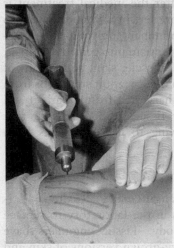

amounts of adipose tissue from various areas of the body. After an incision is made in the skin, the fat is removed through a hollow tube, called a cannula, with the assistance of a powerful vacuum pressure unit that suctions out the fat. The technique can be used as a body-contouring procedure in regions such as the thighs, buttocks, arms, breasts, and abdomen, and to transfer fat to another area of the body. Postsurgical complications

Abdominal Liposuction

that may develop include fat that may enter blood vessels broken during the procedure and obstruct blood flow, infection, loss of feeling in the area, fluid depletion, injury to internal structures, and severe postoperative pain.

◼ Medical Uses of Isotonic, Hypertonic, and Hypotonic Solutions

RBCs and other body cells may be damaged or destroyed if exposed to hypertonic or hypotonic solutions. For this reason, most **intravenous (IV) solutions,** liquids infused into the blood of a vein, are isotonic. Examples are isotonic saline (0.9% NaCl) and D5W, which stands for dextrose 5% in water. Sometimes infusion of a hypertonic solution such as mannitol is useful to treat patients who have *cerebral edema,* excess interstitial fluid in the brain. Infusion of such a solution relieves fluid overload by causing osmosis of water from interstitial fluid into the blood. The kidneys then excrete the excess water from the blood into the urine. Hypotonic solutions, given either orally or through an IV, can be used to treat people who are dehydrated. The water in the hypotonic solution moves from the blood into interstitial fluid and then into body cells to rehydrate them. Water and most sports drinks that you consume to "rehydrate" after a workout are hypotonic relative to your body cells.

◼ Papanicolaou Test

A **Papanicolaou test** (pa-pa-NI-kō-lō), also called a **Pap test** or **Pap smear,** involves collection and microscopic examination of epithelial cells that have been scraped off the apical layer of a tissue. A very common type of Pap test involves examining the cells from the nonkeratinized stratified squamous epithelium of the vagina and cervix (inferior portion) of the uterus. This type of Pap test is performed mainly to detect early changes in the cells of the female reproductive system that may indicate a

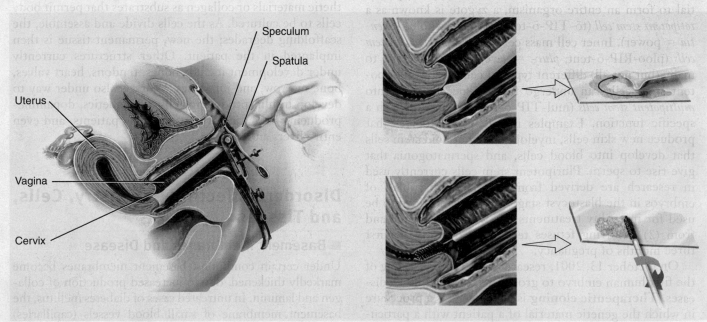

Papanicolaou Test

precancerous condition or cancer. In performing a Pap smear, a physician collects cells, which are then smeared on a microscope slide. The slides are then sent to a laboratory for analysis. Pat tests should be started within three years of the onset of sexual activity, or age 21, whichever comes first. Annual screening is recommended for females ages 21-30 and every 2-3 years for females age 30 or older following three consecutive negative Pap tests.

■ Recombinant DNA

Scientists have developed techniques for inserting genes from other organisms into a variety of host cells. Manipulating the cell in this way can cause the host organism to produce proteins it normally does not synthesize. Organisms so altered are called **recombinants,** and their DNA—a combination of DNA from different sources—is called **recombinant DNA.** When recombinant DNA functions properly, the host will synthesize the protein specified by the new gene it has acquired. The technology that has arisen from the manipulation of genetic material is referred to as **genetic engineering.**

The practical applications of recombinant DNA technology are enormous. Strains of recombinant bacteria now produce large quantities of many important therapeutic substances, including *human growth hormone (hGH)*, required for normal growth and metabolism; *insulin*, a hormone that helps regulate blood glucose level and is used by diabetics; *interferon (IFN)*, an antiviral (and possibly anticancer) substance; and *erythropoietin (EPO)*, a hormone that stimulates production of red blood cells.

■ Stem Cell Research and Therapeutic Cloning

Stem cells are unspecialized cells that have the ability to divide for indefinite periods and give rise to specialized cells. In the context of human development, a zygote (fertilized ovum) is a stem cell. Because it has the potential to form an entire organism, a zygote is known as a *totipotent stem cell* (tō-TIP-ō-tent; *totus-* = whole; *-potentia* = power). Inner cell mass cells, called *pluripotent stem cells* (ploo-RIP-ō-tent; *plur-* = several), can give rise to many (but not all) different types of cells. Later, pluripotent stem cells can undergo further specialization into *multipotent stem cells* (mul-TIP-ō-tent), stem cells with a specific function. Examples include keratinocytes that produce new skin cells, myeloid and lymphoid stem cells that develop into blood cells, and spermatogonia that give rise to sperm. Pluripotent stem cells currently used in research are derived from (1) the embryoblast of embryos in the blastocyst stage that were destined to be used for infertility treatments but were not needed and from (2) nonliving fetuses terminated during the first three months of pregnancy.

On October 13, 2001, researchers reported cloning of the first human embryo to grow cells to treat human diseases. **Therapeutic cloning** is envisioned as a procedure in which the genetic material of a patient with a particular disease is used to create pluripotent stem cells to treat the disease. Using the principles of therapeutic cloning, scientists hope to make an embryo clone of a patient, remove the pluripotent stem cells from the embryo, and then use them to grow tissues to treat particular diseases and disorders, such as cancer, Parkinson disease, Alzheimer disease, spinal cord injury, diabetes, heart disease, stroke, burns, birth defects, osteoarthritis, and rheumatoid arthritis. Presumably, the tissues would not be rejected since they would contain the patient's own genetic material.

Scientists are also investigating the potential clinical applications of *adult stem cells*—stem cells that remain in the body throughout adulthood. Recent experiments suggest that the ovaries of adult mice contain stem cells that can develop into new ova (eggs). If these same types of stem cells are found in the ovaries of adult women, scientists could potentially harvest some of them from a woman about to undergo a sterilizing medical treatment (such as chemotherapy), store them, and then return the stem cells to her ovaries after the medical treatment is completed in order to restore fertility. Studies have also suggested that stem cells in human adult red bone marrow have the ability to differentiate into cells of the liver, kidney, heart, lung, skeletal muscle, skin, and organs of the gastrointestinal tract. In theory, adult stem cells from red bone marrow could be harvested from a patient and then used to repair other tissues and organs in that patient's body without having to use stem cells from embryos.

■ Tissue Engineering

The technology of **tissue engineering** has allowed scientists to grow new tissues in the laboratory to replace damaged tissues in the body. Tissue engineers have already developed laboratory-grown versions of skin and cartilage using scaffolding beds of biodegradable synthetic materials or collagen as substrates that permit body cells to be cultured. As the cells divide and assemble, the scaffolding degrades; the new, permanent tissue is then implanted in the patient. Other structures currently under development include bones, tendons, heart valves, bone marrow, and intestines. Work is also under way to develop insulin-producing cells for diabetics, dopamine-producing cells for Parkinson disease patients, and even entire livers and kidneys. ■

Disorders Affecting Chemistry, Cells, and Tissues

■ Basement Membranes and Disease

Under certain conditions, basement membranes become markedly thickened, due to increased production of collagen and laminin. In untreated cases of diabetes mellitus, the basement membrane of small blood vessels (capillaries) thickens, especially in the eyes and kidneys. Because of this the blood vessels cannot function properly and blindness and kidney failure may result.

■ Cancer

Cancer is a group of diseases characterized by uncontrolled or abnormal cell proliferation. When cells in a part of the body divide without control, the excess tissue that develops is called a **tumor** or **neoplasm** (NĒ-ō-plazm; *neo-* = new). The study of tumors is called **oncology** (on-KOL-ō-jē; *onco-* = swelling or mass). Tumors may be cancerous and often fatal, or they may be harmless. A cancerous neoplasm is called a **malignant tumor** or **malignancy.** One property of most malignant tumors is their ability to undergo **metastasis** (me-TAS-ta-sis), the spread of cancerous cells to other parts of the body. A **benign tumor** is a neoplasm that does not metastasize. An example is a wart. Most benign tumors may be removed surgically if they interfere with normal body function or become disfiguring. Some benign tumors can be inoperable and perhaps fatal.

■ Types of Cancer

The name of a cancer is derived from the type of tissue in which it develops. Most human cancers are **carcinomas** (kar-si-NŌ-maz; *carcin* = cancer; *-omas* = tumors), malignant tumors that arise from epithelial cells. **Melanomas** (mel-a-NŌ-maz; *melan-* = black), for example, are cancerous growths of melanocytes, skin epithelial cells that produce the pigment melanin. **Sarcoma** (sar-KŌ-ma; *sarc-* = flesh) is a general term for any cancer arising from muscle cells or connective tissues. For example, **osteogenic sarcoma** (*osteo-* = bone; *-genic* = origin), the most frequent type of childhood cancer, destroys normal bone tissue. **Leukemia** (loo-KĒ-mē-a; *leuk-* = white; *-emia* = blood) is a cancer of blood-forming organs characterized by rapid growth of abnormal leukocytes (white blood cells). **Lymphoma** (lim-FŌ-ma) is a malignant disease of lymphatic tissue—for example, of lymph nodes.

■ Growth and Spread of Cancer

Cells of malignant tumors duplicate rapidly and continuously. As malignant cells invade surrounding tissues, they often trigger **angiogenesis,** the growth of new networks of blood vessels. Proteins that stimulate angiogenesis in tumors are called **tumor angiogenesis factors (TAFs).** The formation of new blood vessels can occur either by overproduction of TAFs or by the lack of naturally occurring angiogenesis inhibitors. As the cancer grows, it begins to compete with normal tissues for space and nutrients. Eventually, the normal tissue decreases in size and dies. Some malignant cells may detach from the initial (primary) tumor and invade a body cavity or enter the blood or lymph, then circulate to and invade other body tissues, establishing secondary tumors. Malignant cells resist the antitumor defenses of the body. The pain

associated with cancer develops when the tumor presses on nerves or blocks a passageway in an organ so that secretions build up pressure, or as a result of dying tissue or organs.

■ Causes of Cancer

Several factors may trigger a normal cell to lose control and become cancerous. One cause is environmental agents: substances in the air we breathe, the water we drink, and the food we eat. A chemical agent or radiation that produces cancer is called a **carcinogen** (car-SIN-ō-jen). Carcinogens induce **mutations,** permanent changes in the DNA base sequence of a gene. The World Health Organization estimates that carcinogens are associated with 60–90% of all human cancers. Examples of carcinogens are hydrocarbons found in cigarette tar, radon gas from the earth, and ultraviolet (UV) radiation in sunlight.

Intensive research efforts are now directed toward studying cancer-causing genes, or **oncogenes** (ON-kō-jēnz). When inappropriately activated, these genes have the ability to transform a normal cell into a cancerous cell. Most oncogenes derive from normal genes called **proto-oncogenes** that regulate growth and development. The proto-oncogene undergoes some change that causes it either to be expressed inappropriately or to make its products in excessive amounts or at the wrong time. Some oncogenes cause excessive production of growth factors, chemicals that stimulate cell growth. Others may trigger changes in a cell-surface receptor, causing it to send signals as though it were being activated by a growth factor. As a result, the growth pattern of the cell becomes abnormal.

Proto-oncogenes in every cell carry out normal cellular functions until a malignant change occurs. It appears that some proto-oncogenes are activated to oncogenes by mutations in which the DNA of the proto-oncogene is altered. Other proto-oncogenes are activated by a rearrangement of the chromosomes so that segments of DNA are exchanged. Rearrangement activates proto-oncogenes by placing them near genes that enhance their activity.

Some cancers have a viral origin. Viruses are tiny packages of nucleic acids, either RNA or DNA, that can reproduce only while inside the cells they infect. Some viruses, termed **oncogenic viruses,** cause cancer by stimulating abnormal proliferation of cells. For instance, the *human papillomavirus (HPV)* causes virtually all cervical cancers in women. The virus produces a protein that causes proteasomes to destroy p53, a protein that normally suppresses unregulated cell division. In the absence of this suppressor protein, cells proliferate without control.

Recent studies suggest that certain cancers may be linked to a cell having abnormal numbers of chromosomes. As a result, the cell could potentially have extra copies of oncogenes or too few copies of tumor-suppressor genes, which in either case could lead to uncontrolled cell proliferation. There is also some evidence suggesting that cancer may be caused by normal stem cells that develop into cancerous stem cells capable of forming malignant tumors.

Inflammation is a defensive response to tissue damage. It appears that inflammation contributes to various steps in the development of cancer. Some evidence suggests that chronic inflammation stimulates the proliferation of mutated cells and enhances their survival, promotes angiogenesis, and contributes to invasion and metastasis of cancer cells. There is a clear relationship between certain chronic inflammatory conditions and the transformation of inflamed tissue into a malignant tissue. For example, chronic gastritis (inflammation of the stomach lining) and peptic ulcers may be a causative factor in 60-90% of stomach cancers. Chronic hepatitis (inflammation of the liver) and cirrhosis of the liver are believed to be responsible for about 80% of liver cancers. Colorectal cancer is ten time more likely to occur in patients with chronic inflammatory diseases of the colon, such as ulcerative colitis and Crohn's disease. And the relationship between asbestosis and silicosis, two chronic lung inflammatory conditions, and lung cancer has long been recognized. Chronic inflammation is also an underlying contributor to rheumatoid arthritis, Alzheimer disease, depression, schizophrenia, cardiovascular disease, and diabetes.

Carcinogenesis: A Multistep Process

Carcinogenesis (kar′-si-nō-JEN-e-sis), the process by which cancer develops, is a multistep process in which as many as 10 distinct mutations may have to accumulate in a cell before it becomes canc erous. The progression of genetic changes leading to cancer is best understood for colon (colorectal) cancer. Such cancers, as well as lung and breast cancer, take years or decades to develop. In colon cancer, the tumor begins as an area of increased cell proliferation that results from one mutation. This growth then progresses to abnormal, but noncancerous, growths called adenomas. After two or three additional mutations, a mutation of the tumor-suppressor gene p53 occurs and a carcinoma develops. The fact that so many mutations are needed for a cancer to develop indicates that cell growth is normally controlled with many sets of checks and balances. A compromised immune system is also a significant component in carcinogenesis.

Treatment of Cancer

Many cancers are removed surgically. However, when cancer is widely distributed throughout the body or exists in organs such as the brain whose functioning would be greatly harmed by surgery, chemotherapy and radiation therapy may be used instead. Sometimes surgery, chemotherapy, and radiation therapy are used in combination. Chemotherapy involves administering drugs that cause death of cancerous cells. Radiation therapy breaks chromosomes, thus blocking cell division. Because cancerous cells divide rapidly, they are more vulnerable to the destructive effects of chemotherapy and radiation therapy than are normal cells. Unfortunately for the patients, hair follicle cells, red bone marrow cells, and cells lining the gastrointestinal tract also are rapidly dividing. Hence, the side effects of chemotherapy and radiation therapy include hair loss due to death of hair follicle cells, vomiting and nausea due to death of cells lining the stomach and intestines, and susceptibility to infection due to slowed production of white blood cells in red bone marrow.

Treating cancer is difficult because it is not a single disease and because the cells in a single tumor population rarely behave all in the same way. Although most cancers are thought to derive from a single abnormal cell, by the time a tumor reaches a clinically detectable size, it may contain a diverse population of abnormal cells. For example, some cancerous cells metastasize readily, and others do not. Some are sensitive to chemotherapy drugs and some are drug-resistant. Because of differences in drug resistance, a single chemotherapeutic agent may destroy susceptible cells but permit resistant cells to proliferate.

Another potential treatment for cancer that is currently under development is *virotherapy*, the use of viruses to kill cancer cells. The viruses employed in this strategy are designed so that they specifically target cancer cells without affecting the healthy cells of the body. For example, proteins (such as antibodies) that specifically bind to receptors found only in cancer cells are attached to viruses. Once inside the body, the viruses bind to cancer cells and then infect them. The cancer cells are eventually killed once the viruses cause cellular lysis.

Researchers are also investigating the role of *metastasis regulatory genes* that control the ability of cancer cells to undergo metastasis. Scientists hope to develop therapeutic drugs that can manipulate these genes and, therefore, block metastasis of cancer cells.

Marfan Syndrome

Marfan syndrome (MAR-fan) is an inherited disorder caused by a defective fibrillin gene. The result is abnormal development of elastic fibers. Tissues rich in elastic fibers are malformed or weakened. Structures affected most seriously are the covering layer of bones (periosteum), the ligament that suspends the lens of the eye, and the walls of the large arteries. People with Marfan syndrome tend to be tall and have disproportionately long arms, legs, fingers, and toes. A common symptom is blurred vision caused by displacement of the lens of the eye. The most life-threatening complication of Marfan syndrome is weakening of the aorta (the main artery that emerges from the heart), which can suddenly burst.

Progeria and Werner Syndrome

Progeria (prō-JER-ē -a) is a disease characterized by normal development in the first year of life followed by rapid aging. It is caused by a genetic defect in which telomeres are considerably shorter than normal. Symptoms include dry and wrinkled skin, total baldness, and birdlike facial features. Death usually occurs around age 13.

Werner syndrome is a rare, inherited disease that causes a rapid acceleration of aging, usually while the person is only in his or her twenties. It is characterized by wrinkling of the skin, graying of the hair and baldness, cataracts, muscular atrophy, and a tendency to develop diabetes mellitus, cancer, and cardiovascular disease. Most afflicted individuals die before age 50. Recently, the gene that causes Werner syndrome has been identified. Researchers hope to use the information to gain insight into the mechanisms of aging, as well as to help those suffering from the disorder.

Sjögren's Syndrome

Sjögren's syndrome (SHŌ -grenz) is a common autoimmune disorder that causes inflammation and destruction of exocrine glands, especially the lacrimal (tear) glands and salivary glands. Signs include dryness of the eyes, mouth, nose, ears, skin, and vagina, and salivary gland enlargement. Systemic effects include fatigue, arthritis, difficulty in swallowing, pancreatitis (inflammation of the pancreas), pleuritis (inflammation of the pleurae of the lungs), and muscle and joint pain. The disorder affects females more than males by a ratio of 9 to 1. About 20% of older adults experience some signs of Sjögren's. Treatment is supportive, including using artificial tears to moisten the eyes, sipping fluids, chewing sugarless gum, using a saliva substitute to moisten the mouth, and using moisturizing creams for the skin. If symptoms or complications are sever, medications may be used. These include cyclosporine eyedrops, polocarpine to increase saliva production, immunosuppressants, nonsteroidal anti-inflammatory drugs, and corticosteroids.

Systemic Lupus Erythematosus

Systemic lupus erythematosus (er-i-thē -ma-TŌ -sus), **SLE,** or simply lupus, is a chronic inflammatory disease of connective tissue occurring mostly in nonwhite women during their childbearing years. It is an autoimmune disease that can cause tissue damage in every body system. The disease, which can range from a mild condition in most patients to a rapidly fatal disease, is marked by periods of exacerbation and remission. The prevalence of SLE is about 1 in 2000, with females more likely to be afflicted than males by a ratio of 8 or 9 to 1.

Although the cause of SLE is unknown, genetic, environmental, and hormonal factors all have been implicated. The genetic component is suggested by studies of twins and family history. Environmental factors include viruses, bacteria, chemicals, drugs, exposure to excessive sunlight, and emotional stress. Sex hormones, such as estrogens, may also trigger SLE.

Signs and symptoms of SLE include painful joints, low-grade fever, fatigue, mouth ulcers, weight loss, enlarged lymph nodes and spleen, sensitivity to sunlight, rapid loss of large amounts of scalp hair, and anorexia. A distinguishing feature of lupus is an eruption across the bridge of the nose and cheeks called a "butterfly rash." Other skin lesions may occur, including blistering and ulceration. The erosive nature of some SLE skin lesions was thought to resemble the damage inflicted by the bite of a wolf—thus, the name *lupus* (= wolf). The most serious complications of the disease involve inflammation of the kidneys, liver, spleen, lungs, heart, brain, and gastrointestinal tract. Because there is no cure for SLE, treatment is supportive, including anti-inflammatory drugs, such as aspirin, and immunosuppressive drugs.

Tay-Sachs Disease

Some disorders are caused by faulty or absent lysosomal enzymes. For instance, **Tay-Sachs disease,** which most often affects children of Ashkenazi (eastern European Jewish) descent, is an inherited condition characterized by the absence of a single lysosomal enzyme called Hex A. This enzyme normally breaks down a membrane glycolipid called ganglioside G_{M2} that is especially prevalent in nerve cells. As the excess ganglioside G_{M2} accumulates, the nerve cells function less efficiently. Children with Tay-Sachs disease typically experience seizures and muscle rigidity. They gradually become blind, demented, and uncoordinated and usually die before the age of 5. Tests can now reveal whether an adult is a carrier of the defective gene. ∎

Additional Clinical Considerations

Adhesions

Scar tissue can form **adhesions,** abnormal joining of tissues. Adhesions commonly form in the abdomen around a site of previous inflammation such as an inflamed appendix, and they can develop after surgery. Although adhesions do not always cause problems, they can decrease tissue flexibility, cause obstruction (such as in the intestine), and make a subsequent operation more difficult. An *adhesiotomy,* the surgical release of adhesions, may be required.

Chondroitin Sulfate, Glucosamine, and Joint Disease

In recent years, chondroitin sulfate and glucosamine (a proteoglycan) have been used as nutritional supplements either alone or in a combination to promote and maintain the structure and function of joint cartilage, to provide pain relief from osteoarthritis, and to reduce joint inflammation. Although these supplements have benefited some individuals with moderate to severe osteoarthritis, the benefit is minimal in lesser cases. More research is needed to determine how they act and why they help some people and not others.

■ Digitalis increases Ca²⁺ in Heart Muscle Cells

Digitalis often is given to patients with *heart failure*, a condition of weakened pumping action by the heart. Digitalis exerts its effect by slowing the action of the sodium-potassium pumps, which lets more Na^+ accumulate inside heart muscle cells. The result is a decreased Na^+ concentration gradient across the plasma membrane, which causes the Na^+/Ca^{2+} antiporters to slow down. As a result, more Ca^{2+} remains inside heart muscle cells. The slight increase in the level of Ca^{2+} in the cytosol of heart muscle cells increases the force of their contractions and thus strengthens the force of the heartbeat.

■ Fatty Acids in Health and Disease

As its name implies, a group of fatty acids called **essential fatty acids (EFAs)** is essential to human health. However, they cannot be made by the human body and must be obtained from foods or supplements. Among the more important EFAs are *omega-3 fatty acids*, *omega-6 fatty acids*, and cis-*fatty acids*.

Omega-3 and omega-6 fatty acids are polyunsaturated fatty acids that are believed to work together to promote health. They may have a protective effect against heart disease and stroke by lowering total cholesterol, raising HDL (high-density lipoproteins or "good cholesterol") and lowering LDL (low-density lipoproteins or "bad cholesterol"). In addition, omega-3 and omega-6 fatty acids decrease bone loss by increasing calcium utilization by the body; reduce symptoms of arthritis due to inflammation; promote wound healing; improve certain skin disorders (psoriasis, eczema, and acne); and improve mental functions. Primary sources of omega-3 fatty acids include flaxseed, fatty fish, oils that have large amounts of polyunsaturated fatty acids, fish oils, and walnuts. Primary sources of omega-6 fatty acids include most processed foods (cereals, breads, white rice), eggs, baked goods, oils with large amounts of polyunsaturated fatty acids, and meats (especially organ meats, such as liver).

The hydrogen atoms on either side of the double bond in oleic acid are on the same side of the unsaturated fatty acid. Such an unsaturated fatty acid is called a *cis*-fatty acid. *Cis*-fatty acids are nutritionally beneficial unsaturated fatty acids that are used by the body to produce hormone-like regulators and cell membranes. However, when *cis*-fatty acids are heated, pressurized, and combined with a catalyst (usually nickel) in a process called *hydrogenation*, they are changed to unhealthy *trans*-fatty acids. In *trans*-fatty acids the hydrogen atoms are on opposite sides of the double bond of an unsaturated fatty acid. Hydrogenation is used by manufacturers to make vegetable oils solid at room temperature and less likely to turn rancid. Hydrogenated or *trans*-fatty acids are common in commercially baked goods (crackers, cakes, and cookies), salty snack foods, some margarines, and fried foods (donuts and french fries). When oil is used for frying and if the oil is reused (like in fast food French fry machines) *cis*-fatty acids are converted to *trans*-fatty acids. If a product label contains the words hydrogenated or partially hydrogenated, then the product contains *trans*-fatty acids. Among the adverse effects of *trans*-fatty acids are an increase in total cholesterol, a decrease in HDL, an increase in LDL, and an increase in triglycerides. These effects, which can increase the risk of heart disease and other cardiovascular diseases, are similar to those caused by saturated fats.

■ Free Radicals and Their Effects on Health

In our bodies, several processes can generate free radicals, including exposure to ultraviolet radiation in sunlight, exposure to x-rays, and some reactions that occur during normal metabolic processes. Certain harmful substances, such as carbon tetrachloride (a solvent used in dry cleaning), also give rise to free radicals when they participate in metabolic reactions in the body. Among the many disorders, diseases, and conditions linked to oxygen-derived free radicals are cancer, atherosclerosis, Alzheimer disease, emphysema, diabetes mellitus, cataracts, macular degeneration, rheumatoid arthritis, and deterioration associated with aging. Consuming more *antioxidants*—substances that inactivate oxygen-derived free radicals—is thought to slow the pace of damage caused by free radicals. Important dietary antioxidants include selenium, zinc, beta-carotene, and vitamins C and E.

■ Genomics

In the last decade of the twentieth century, the genomes of humans, mice, fruit flies, and more than 50 microbes were sequenced. As a result, research in the field of **genomics,** the study of the relationships between the genome and the biological functions of an organism, has flourished. The Human Genome Project began in June 1990 as an effort to sequence all of the nearly 3.2 billion nucleotides of our genome, and was completed in April 2003. More than 99.9% of the nucleotide bases are identical in everyone. Less than 0.1% of our DNA (1 in each 1000 bases) accounts for inherited differences among humans. Surprisingly, at least half of the human genome consists of repeated sequences that do not code for proteins, so-called "junk" DNA. The average gene consists of 3000 nucleotides, but sizes vary greatly. The largest known human gene, with 2.4 million nucleotides, codes for the protein dystrophin. Scientists now know that the total number of genes in the human genome is about 30,000,

Human Genome Project: logo with DNA chromosome

far fewer than the 100,000 previously predicted to exist. Information regarding the human genome and how it is affected by the environment seeks to identify and discover the functions of the specific genes that play a role in genetic diseases. Genomic medicine also aims to design new drugs and to provide screening tests to enable physicians to provide more effective counseling and treatment for disorders with significant genetic components such as hypertension (high blood pressure), obesity, diabetes, and cancer.

■ Harmful and Beneficial Effects of Radiation

Radioactive isotopes may have either harmful or helpful effects. Their radiations can break apart molecules, posing a serious threat to the human body by producing tissue damage and/or causing various types of cancer. Although the decay of naturally occurring radioactive isotopes typically releases just a small amount of radiation into the environment, localized accumulations can occur. Radon-222, a colorless and odorless gas that is a naturally occurring radioactive breakdown product of uranium, may seep out of the soil and accumulate in buildings. It is not only associated with many cases of lung cancer in smokers but has also been implicated in many cases of lung cancer in nonsmokers. Beneficial effects of certain radioisotopes include their use in medical imaging procedures to diagnose and treat certain disorders. Some radioisotopes can be used as **tracers** to follow the movement of certain substances through the body. Thallium-201 is used to monitor blood flow through the heart during an exercise stress test. Iodine-131 is used to detect cancer of the thyroid gland and to assess its size and activity, and may also be used to destroy part of an overactive thyroid gland. Cesium-137 is used to treat advanced cervical cancer, and iridium is used to treat prostate cancer.

■ Smooth ER and Drug Tolerance

One of the functions of smooth ER, as noted earlier, is to detoxify certain drugs. Individuals who repeatedly take such drugs, such as the sedative phenobarbital, develop changes in the smooth ER in their liver cells. Prolonged administration of phenobarbital results in increased tolerance to the drug; the same dose no longer produces the same degree of sedation. With repeated exposure to the drug, the amount of smooth ER and its enzymes increases to protect the cell from its toxic effects. As the amount of smooth ER increases, higher and higher dosages of the drug are needed to achieve the original effect.

■ Tumor-Suppressor Genes

Abnormalities in genes that regulate the cell cycle or apoptosis are associated with many diseases. For example, damage to genes called **tumor-suppressor genes,** which produce proteins that normally inhibit cell division, causes some types of cancer. Loss or alteration of a tumor-suppressor gene called *p53* on chromosome 17 is the most

common genetic change leading to a wide variety of tumors, including breast and colon cancers. The normal p53 protein arrests cells in the G_1 phase, which prevents cell division. Normal p53 protein also assists in repair of damaged DNA and induces apoptosis in the cells where DNA repair was not successful. For this reason, the p53 gene is nicknamed "the guardian angel of the genome."

■ Viruses and Receptor-Mediated Endocytosis

Although receptor-mediated endocytosis normally imports needed materials, some viruses are able to use this mechanism to enter and infect body cells. For example, the human immunodeficiency virus (HIV), which causes acquired immunodeficiency syndrome (AIDS), can attach to a receptor called CD4. This receptor is present in the plasma membrane of white blood cells called helper T cells. After binding to CD4, HIV enters the helper T cell via receptor-mediated endocytosis.

Medical Terminology

Anaplasia (an′-a-PLĀ-zē -a; *an-* = not; *-plasia* = to shape) The loss of tissue differentiation and function that is characteristic of most malignancies.

Atrophy (AT-rō -fē ; *a-* = without; *-trophy* = nourishment) A decrease in the size of cells, with a subsequent decrease in the size of the affected tissue or organ; wasting away.

Biopsy (BĪ-op-sē ; *bio-*=life; *-opsy*=to view) The removal of a sample of living tissue for microscopic examination to help diagnose disease.

Dysplasia (dis-PLĀ-zē -a; *dys-* = abnormal) Alteration in the size, shape, and organization of cells due to chronic irritation or inflammation; may progress to neoplasia (tumor formation, usually malignant) or revert to normal if the irritation is removed.

Hyperplasia (hī -per-PLĀ-zē-a; *hyper-* = over) Increase in the number of cells of a tissue due to an increase in the frequency of cell division.

Hypertrophy (hī -PER-trō -fē) Increase in the size of a tissue because its cells enlarge without undergoing cell division.

Metaplasia (met′-a-PLĀ-zē-a; *meta-* = change) The transformation of one type of cell into another.

Progeny (PROJ-e-nē; *pro-* = forward; *-geny* = production) Offspring or descendants.

Proteomics (prō′-tē-Ō-miks; *proteo-* = protein) The study of the proteome (all of an organism's proteins) in order to identify all the proteins produced; it involves determining the three-dimensional structure of proteins so that drugs can be designed to alter protein activity to help in the treatment and diagnosis of disease.

Tissue rejection An immune response of the body directed at foreign proteins in a transplanted tissue or organ;

immunosuppressive drugs, such as cyclosporine, have largely over-come tissue rejection in heart-, kidney-, and liver-transplant patients.

Tissue transplantation   The replacement of a diseased or injured tissue or organ. The most successful transplants involve use of a person's own tissues or those from an identical twin.

Tumor marker   A substance introduced into circulation by tumor cells that indicates the presence of a tumor, as well as the specific type. Tumor markers may be used to screen, diagnose, make a prognosis, evaluate a response to treatment, and monitor for recurrence of cancer.

Xenotransplantation (zen′-ō-trans-plan-TĀ-shun; *xeno-* = strange, foreign)   The replacement of a diseased or injured tissue or organ with cells or tissues from an animal. Porcine (from pigs) and bovine (from cows) heart valves are used for some heart-valve replacement surgeries.

Chapter 3 | The Integumentary System

The integumentary system is composed of the skin, hair, oil and sweat glands, nails, and sensory receptors. The integumentary system contributes to homeostasis by protecting the body and helping regulate body temperature. Of all the body's organs, none is more easily inspected or more exposed to infection, disease, and injury than the skin. **Dermatology** is the medical specialty that deals with the diagnosis and treatment of integumentary system disorders.

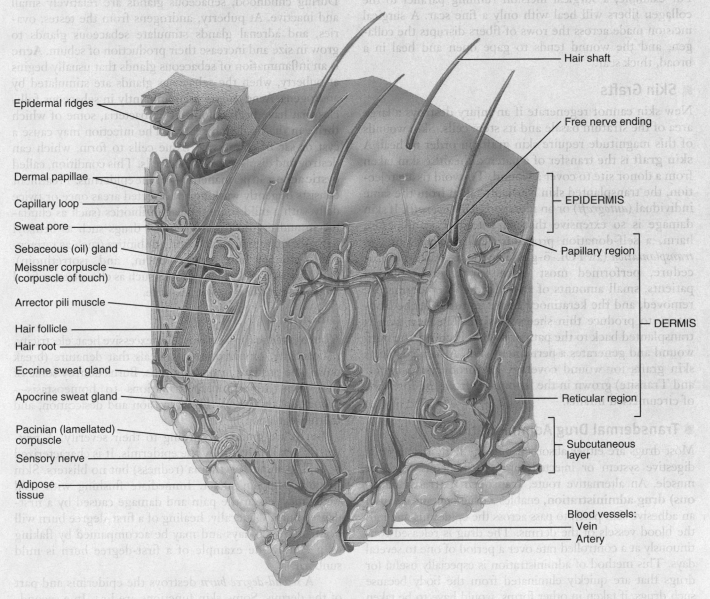

- Hair shaft
- Epidermal ridges
- Free nerve ending
- Dermal papillae
- Capillary loop
- EPIDERMIS
- Sweat pore
- Papillary region
- Sebaceous (oil) gland
- Meissner corpuscle (corpuscle of touch)
- Arrector pili muscle
- DERMIS
- Hair follicle
- Hair root
- Eccrine sweat gland
- Apocrine sweat gland
- Reticular region
- Pacinian (lamellated) corpuscle
- Subcutaneous layer
- Sensory nerve
- Adipose tissue
- Blood vessels:
  - Vein
  - Artery

Sectional view of skin and subcutaneous layer

Tests, Procedures, And Techniques

■ Lines of Cleavage and Surgery

In certain regions of the body, collagen fibers tend to orient more in one direction than another. **Lines of cleavage (tension lines)** in the skin indicate the predominant direction of underlying collagen fibers. The lines are especially evident on the palmar surfaces of the fingers, where they are aligned with the long axis of the digits. Knowledge of lines of cleavage is especially important to plastic surgeons. For example, a surgical incision running parallel to the collagen fibers will heal with only a fine scar. A surgical incision made across the rows of fibers disrupts the collagen, and the wound tends to gape open and heal in a broad, thick scar.

■ Skin Grafts

New skin cannot regenerate if an injury destroys a large area of the stratum basale and its stem cells. Skin wounds of this magnitude require skin grafts in order to heal. A **skin graft** is the transfer of a patch of healthy skin taken from a donor site to cover a wound. To avoid tissue rejection, the transplanted skin is usually taken from the same individual (*autograft*) or an identical twin (*isograft*). If skin damage is so extensive that an autograft would cause harm, a self-donation procedure called *autologous skin transplantation* (awTOL-ō-gus) may be used. In this procedure, performed most often for severely burned patients, small amounts of an individual's epidermis are removed, and the keratinocytes are cultured in the laboratory to produce thin sheets of skin. The new skin is transplanted back to the patient so that it covers the burn wound and generates a permanent skin. Also available as skin grafts for wound coverage are products (Apligraft and Transite) grown in the laboratory from the foreskins of circumcised infants.

■ Transdermal Drug Administration

Most drugs are either absorbed into the body through the digestive system or injected into subcutaneous tissue or muscle. An alternative route, **transdermal (transcutaneous) drug administration,** enables a drug contained within an adhesive skin patch to pass across the epidermis and into the blood vessels of the dermis. The drug is released continuously at a controlled rate over a period of one to several days. This method of administration is especially useful for drugs that are quickly eliminated from the body because such drugs, if taken in other forms, would have to be taken quite frequently. Because the major barrier to penetration of most drugs is the stratum corneum, transdermal absorption is most rapid in regions of the skin where this layer is thin, such as the scrotum, face, and scalp. A growing number of drugs are available for transdermal administration, including nitroglycerin, for prevention of angina pectoris (chest pain associated with heart disease); scopolamine, for motion sickness; estradiol, used for estrogenreplacement therapy during menopause; ethinyl estradiol and norel-gestromin in contraceptive patches; nicotine, used to help people stop smoking; and fentanyl, used to relieve severe pain in cancer patients. ■

Disorders Affecting The Integumentary System

■ Acne

During childhood, sebaceous glands are relatively small and inactive. At puberty, androgens from the testes, ovaries, and adrenal glands stimulate sebaceous glands to grow in size and increase their production of sebum. **Acne** is an inflammation of sebaceous glands that usually begins at puberty, when the sebaceous glands are stimulated by androgens. Acne occurs predominantly in sebaceous follicles that have been colonized by bacteria, some of which thrive in the lipid-rich sebum. The infection may cause a cyst or sac of connective tissue cells to form, which can destroy and displace epidermal cells. This condition, called **cystic acne,** can permanently scar the epidermis. Treatment consists of gently washing the affected areas once or twice daily with a mild soap, topical antibiotics (such as clindamycin and erythromycin), topical drugs such as benzoyl peroxide or tretinoin, and oral antibiotics (such as tetracycline, minocycline, erythromycin, and isotretinoin). Contrary to popular belief, foods such as chocolate or fried foods do not cause or worsen acne.

■ Burns

A **burn** is tissue damage caused by excessive heat, electricity, radioactivity, or corrosive chemicals that denature (break down) the proteins in the skin cells. Burns destroy some of the skin's important contributions to homeostasis—protection against microbial invasion and desiccation, and thermoregulation.

Burns are graded according to their severity. A *first-degree burn* involves only the epidermis. It is characterized by mild pain and erythema (redness) but no blisters. Skin functions remain intact. Immediate flushing with cold water may lessen the pain and damage caused by a first-degree burn. Generally, healing of a first-degree burn will occur in 3 to 6 days and may be accompanied by flaking or peeling. One example of a first-degree burn is mild sunburn.

A *second-degree burn* destroys the epidermis and part of the dermis. Some skin functions are lost. In a second-degree burn, redness, blister formation, edema, and pain result. In a blister the epidermis separates from the dermis due to the accumulation of tissue fluid between them. Associated structures, such as hair follicles, sebaceous glands, and sweat glands, usually are not injured. If there is no infection, second-degree burns heal without skin grafting in about 3 to 4 weeks, but scarring may result. First- and second-degree burns are collectively referred to as *partial-thickness burns.*

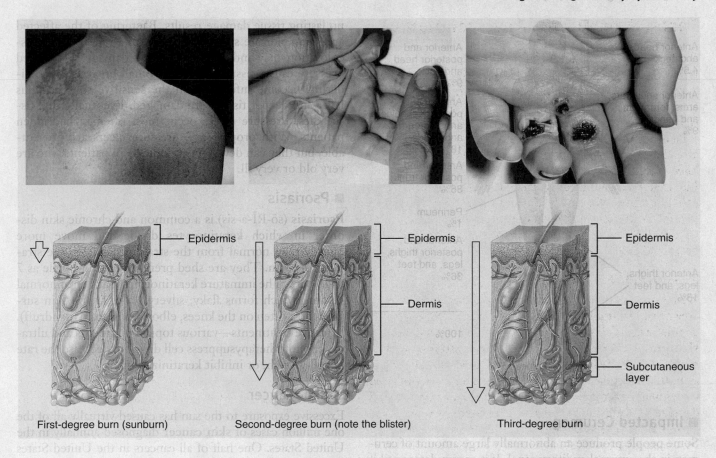

First-degree burn (sunburn)   Second-degree burn (note the blister)   Third-degree burn

A *third-degree burn* or *full-thickness burn* destroys the epidermis, dermis, and subcutaneous layer. Most skin functions are lost. Such burns vary in appearance from marble-white to mahogany colored to charred, dry wounds. There is marked edema, and the burned region is numb because sensory nerve endings have been destroyed. Regeneration occurs slowly, and much granulation tissue forms before being covered by epithelium. Skin grafting may be required to promote healing and to minimize scarring.

The injury to the skin tissues directly in contact with the damaging agent is the *local effect* of a burn. Generally, however, the *systemic effects* of a major burn are a greater threat to life. The systemic effects of a burn may include (1) a large loss of water, plasma, and plasma proteins, which causes shock; (2) bacterial infection; (3) reduced circulation of blood; (4) decreased production of urine; and (5) diminished immune responses.

The seriousness of a burn is determined by its depth and extent of area involved, as well as the person's age and general health. According to the American Burn Association's classification of burn injury, a major burn includes third-degree burns over 10% of body surface area; or second-degree burns over 25% of body surface area; or any third-degree burns on the face, hands, feet, or *perineum* (per-i-NĒ-um, which includes the anal and urogenital regions). When the burn area exceeds 70%, more than half the victims die. A quick means for estimating the surface area affected by a burn in an adult is the **rule of nines**:

1. Count 9% if both the anterior and posterior surfaces of the head and neck are affected.
2. Count 9% for both the anterior and posterior surfaces of each upper limb (total of 18% for both upper limbs).
3. Count four times nine or 36% for both the anterior and posterior surfaces of the trunk, including the buttocks.
4. Count 9% for the anterior and 9% for the posterior surfaces of each lower limb as far up as the buttocks (total of 36% for both lower limbs).
5. Count 1% for the perineum.

Many people who have been burned in fires also inhale smoke. If the smoke is unusually hot or dense or if inhalation is prolonged, serious problems can develop. The hot smoke can damage the trachea (windpipe), causing its lining to swell. As the swelling narrows the trachea, airflow into the lungs is obstructed. Further, small airways inside the lungs can also narrow, producing wheezing or shortness of breath. A person who has inhaled smoke is given oxygen through a face mask, and a tube may be inserted into the trachea to assist breathing.

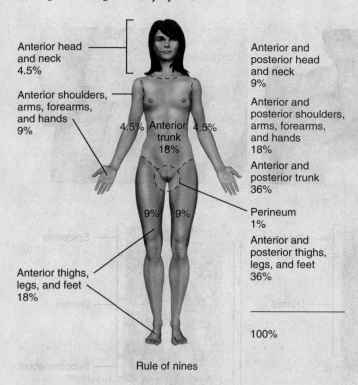

Anterior head and neck 4.5%

Anterior shoulders, arms, forearms, and hands 9%

4.5% Anterior trunk 18% 4.5%

Anterior thighs, legs, and feet 18%

9% 9%

Anterior and posterior head and neck 9%

Anterior and posterior shoulders, arms, forearms, and hands 18%

Anterior and posterior trunk 36%

Perineum 1%

Anterior and posterior thighs, legs, and feet 36%

100%

Rule of nines

■ Impacted Cerumen

Some people produce an abnormally large amount of cerumen in the external auditory canal. If it accumulates until it becomes impacted (firmly wedged), sound waves may be prevented from reaching the eardrum. Treatments for **impacted cerumen** include periodic ear irrigation with enzymes to dissolve the wax and removal of wax with a blunt instrument by trained medical personnel. The use of cotton-tipped swabs or sharp objects is not recommended for this purpose because they may push the cerumen further into the external auditory canal and damage the eardrum.

■ Pressure Ulcers

Pressure ulcers, also known as *decubitus ulcers* (dē-KŪ-bi-tus) or *bedsores*, are caused by a constant deficiency of blood flow to tissues. Typically the affected tissue overlies a bony projection that has been subjected to prolonged pressure against an object such as a bed, cast, or splint. If the pressure is relieved in a few hours, redness occurs but

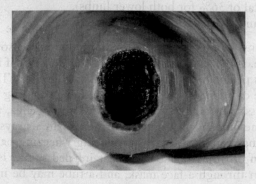

Pressure ulcer on heel

no lasting tissue damage results. Blistering of the affected area may indicate superficial damage; a reddish-blue discoloration may indicate deep tissue damage. Prolonged pressure causes tissue ulceration. Small breaks in the epidermis become infected, and the sensitive subcutaneous layer and deeper tissues are damaged. Eventually, the tissue dies. Pressure ulcers occur most often in bedridden patients. With proper care, pressure ulcers are preventable, but they can develop very quickly in patients who are very old or very ill.

■ Psoriasis

Psoriasis (sō-RĪ-a-sis) is a common and chronic skin disorder in which keratinocytes divide and move more quickly than normal from the stratum basale to the stratum corneum. They are shed prematurely in as little as 7 to 10 days. The immature keratinocytes make an abnormal keratin, which forms flaky, silvery scales at the skin surface, most often on the knees, elbows, and scalp (dandruff). Effective treatments—various topical ointments and ultraviolet phototherapysuppress cell division, decrease the rate of cell growth, or inhibit keratinization.

■ Skin Cancer

Excessive exposure to the sun has caused virtually all of the one million cases of **skin cancer** diagnosed annually in the United States. One half of all cancers in the United States are skin cancers. There are three common forms of skin cancer. **Basal cell carcinomas** account for about 78% of all skin cancers. The tumors arise from cells in the stratum basale of the epidermis and rarely metastasize. **Squamous cell carcinomas,** which account for about 20% of all skin cancers, arise from squamous cells of the epidermis, and they have a variable tendency to metastasize. Most arise from preexisting lesions of damaged tissue on sun-exposed skin. Basal and squamous cell carcinomas are together known as *nonmelanoma skin cancer.* They are 50% more common in males than in females.

    Malignant melanomas arise from melanocytes and account for about 2% of all skin cancers. The estimated lifetime risk of developing melanoma is now 1 in 75, double the risk only 20 years ago. In part, this increase is due to depletion of the ozone layer, which absorbs some UV light high in the atmosphere. But the main reason for the increase is that more people are spending more time in the sun and in tanning beds. Malignant melanomas metastasize rapidly and can kill a person within months of diagnosis.

    The key to successful treatment of malignant melanoma is early detection. The early warning signs of malignant melanoma are identified by the acronym ABCD. *A* is for *asymmetry;* malignant melanomas tend to lack symmetry. This means that they have irregular shapes, such as two very different looking halves. *B* is for *border;* malignant melanomas have irregular—notched, indented, scalloped, or indistinct—borders. *C* is for *color;* malignant melanomas have uneven coloration and may contain several colors. *D*

Normal nevus (mole)   Malignant melanoma

is for *diameter*; ordinary moles typically are smaller than 6 mm (0.25 in.), about the size of a pencil eraser. Once a malignant melanoma has the characteristics of A, B, and C, it is usually larger than 6 mm.

Among the risk factors for skin cancer are the following:

1. *Skin type.* Individuals with light-colored skin who never tan but always burn are at high risk.
2. *Sun exposure.* People who live in areas with many days of sunlight per year and at high altitudes (where ultraviolet light is more intense) have a higher risk of developing skin cancer. Likewise, people who engage in outdoor occupations and those who have suffered three or more severe sunburns have a higher risk.
3. *Family history.* Skin cancer rates are higher in some families than in others.
4. *Age.* Older people are more prone to skin cancer owing to longer total exposure to sunlight.
5. *Immunological status.* Immunosuppressed individuals have a higher incidence of skin cancer.

Additional Clinical Considerations

◼ Chemotherapy and Hair Loss

Chemotherapy is the treatment of disease, usually cancer, by means of chemical substances or drugs. Chemotherapeutic agents interrupt the life cycle of rapidly dividing cancer cells. Unfortunately, the drugs also affect other rapidly dividing cells in the body, such as the hair matrix cells of a hair. It is for this reason that individuals undergoing chemotherapy experience hair loss. Since about 15% of the hair matrix cells of scalp hairs are in the resting stage, these cells are not affected by chemotherapy. Once chemotherapy is stopped, the hair matrix cells replace lost hair follicles and hair growth resumes.

◼ Hair and Hormones

At puberty, when the testes begin secreting significant quantities of androgens (masculinizing sex hormones), males develop the typical male pattern of hair growth throughout the body, including a beard and a hairy chest. In females at puberty, the ovaries and the adrenal glands produce small quantities of androgens, which promote hair growth throughout the body including the axillae and pubic region. Occasionally, a tumor of the adrenal glands,

testes, or ovaries produces an excessive amount of androgens. The result in females or prepubertal males is **hirsutism** (HER-soo-tizm; *hirsut-* = shaggy), a condition of excessive body hair.

Surprisingly, androgens also must be present for occurrence of the most common form of baldness, **androgenic alopecia** or **male-pattern baldness.** In genetically predisposed adults, androgens inhibit hair growth. In men, hair loss usually begins with a receding hairline followed by hair loss in the temples and crown. Women are more likely to have thinning of hair on top of the head. The first drug approved for enhancing scalp hair growth was minoxidil (Rogaine®). It causes vasodilation (widening of blood vessels), thus increasing circulation. In about a third of the people who try it, minoxidil improves hair growth, causing scalp follicles to enlarge and lengthening the growth cycle. For many, however, the hair growth is meager. Minoxidil does not help people who already are bald.

◼ Hair Removal

A substance that removes hair is called a **depilatory.** It dissolves the protein in the hair shaft, turning it into a gelatinous mass that can be wiped away. Because the hair root is not affected, regrowth of the hair occurs. In **electrolysis,** an electric current is used to destroy the hair matrix so the hair cannot regrow. **Laser treatments** may also be used to remove hair.

◼ Skin Color as a Diagnostic Clue

The color of skin and mucous membranes can provide clues for diagnosing certain conditions. When blood is not picking up an adequate amount of oxygen from the lungs, as in someone who has stopped breathing, the mucous membranes, nail beds, and skin appear bluish or **cyanotic** (sī-a-NOT-ik; *cyan-* = blue). **Jaundice** (JON-dis; *jaund-* = yellow) is due to a buildup of the yellow pigment bilirubin in the skin. This condition gives a yellowish appearance to the skin and the whites of the eyes, and usually indicates liver disease. **Erythema** (er-e-THĒ-ma; *eryth-* = red), redness of the skin, is caused by engorgement of capillaries in the dermis with blood due to skin injury, exposure to heat, infection, inflammation, or allergic reactions. **Pallor** (PAL-or), or paleness of the skin, may occur in conditions such as shock and anemia. All skin color changes are observed most readily in people with lighter-colored skin and may be more difficult to discern in people with darker skin. However, examination of the nail beds and gums can provide some information about circulation in individuals with darker skin.

◼ Sun Damage, Sunscreens, and Sunblocks

Although basking in the warmth of the sun may feel good, it is not a healthy practice. There are two forms of ultraviolet radiation that affect the health of the skin. Longer-wavelength ultraviolet A (UVA) rays make up nearly 95% of the ultraviolet radiation that reaches the earth. UVA rays are not absorbed by the ozone layer. They penetrate the furthest into the skin, where they are absorbed by melanocytes and thus are involved in sun tanning. UVA rays also depress the immune system. Shorter-wavelength ultraviolet B (UVB) rays are partially absorbed by the ozone layer and do not penetrate the skin as deeply as UVA rays. UVB rays cause sunburn and are responsible for most of the tissue damage (production of oxygen free radicals that disrupt collagen and elastic fibers) that results in wrinkling and aging of the skin and cataract formation.

Both UVA and UVB rays are thought to cause skin cancer. Long-term overexposure to sunlight results in dilated blood vessels, age spots, freckles, and changes in skin texture.

Exposure to ultraviolet radiation (either natural sunlight or the artificial light of a tanning booth) may also produce **photosensitivity,** a heightened reaction of the skin after consumption of certain medications or contact with certain substances. Photosensitivity is characterized by redness, itching, blistering, peeling, hives, and even shock. Among the medications or substances that may cause a photosensitivity reaction are certain antibiotics (tetracycline), nonsteroidal anti-inflammatory drugs (ibuprofen or naproxen), certain herbal supplements (St. John's Wort), some birth control pills, some high blood pressure medications, some antihistamines, and certain artificial sweeteners, perfumes, after shaves, lotions, detergents, and medicated cosmetics.

Self-tanning lotions (sunless tanners), topically applied substances, contain a color additive (dihydroxyacetone) that produces a tanned appearance by interacting with proteins in the skin.

Sunscreens are topically applied preparations that contain various chemical agents (such as benzophenone or one of its derivatives) that absorb UVB rays, but let most of the UVA rays pass through.

Sunblocks are topically applied preparations that contain substances such as zinc oxide that reflect and scatter both UVB and UVA rays.

Both sunscreens and sunblocks are graded according to a *sun protection factor (SPF)* rating, which measures the level of protection they supposedly provide against UV rays. The higher the rating, presumably the greater the degree of protection. As a precautionary measure, individuals who plan to spend a significant amount of time in the sun should use a sunscreen or a sunblock with an SPF of 15 or higher. Although sunscreens protect against sunburn, there is considerable debate as to whether they actually protect against skin cancer. In fact, some studies suggest that sunscreens increase the incidence of skin cancer because of the false sense of security they provide.

Medical Terminology

Abrasion (a-BRĀ-shun; *ab-* = away; *-rasion* = scraped) An area where skin has been scraped away.

Athlete's foot A superficial fungal infection of the skin of the foot.

Blister A collection of serous fluid within the epidermis or between the epidermis and dermis, due to short-term but severe friction. The term **bulla** (BUL-a) refers to a large blister.

Callus (KAL-lus = hardskin) An area of hardened and thickened skin that is usually seen in palms and soles and is due to persistent pressure and friction.

Cold sore A lesion, usually in oral mucous membrane, caused by Type 1 herpes simplex virus (HSV) transmitted by oral or respiratory routes. The virus remains dormant until triggered by factors such as ultraviolet light, hormonal changes, and emotional stress. Also called a **fever blister.**

Comedo (KOM-ē-dō; *comedo* = to eat up) A collection of sebaceous material and dead cells in the hair follicle and excretory duct of the sebaceous (oil) gland. Usually found over the face, chest, and back, and more commonly during adolescence. Also called a blackhead.

Contact dermatitis (der-ma-TĪ-tis; *dermat-* = skin; *-itis* = inflammation of) Inflammation of the skin characterized by redness, itching, and swelling and caused by exposure of the skin to chemicals that bring about an allergic reaction, such as poison ivy toxin.

Corn A painful conical thickening of the stratum corneum of the epidermis found principally over toe joints and between the toes, often caused by friction or pressure. Corns may be hard or soft, depending on their location. Hard corns are usually found over toe joints, and soft corns are usually found between the fourth and fifth toes.

Cyst (SIST; *cyst* = sac containing fluid) A sac with a distinct connective tissue wall, containing a fluid or other material.

Eczema (EK-ze-ma; *ekzeo-* = to boil over) An inflammation of the skin characterized by patches of red, blistering, dry, extremely itchy skin. It occurs mostly in skin creases in the wrists, backs of the knees, and fronts of the elbows. It typically begins in infancy and many children outgrow the condition. The cause is unknown but is linked to genetics and allergies.

Frostbite Local destruction of skin and subcutaneous tissue on exposed surfaces as a result of extreme cold. In mild cases, the skin is blue and swollen and there is slight pain. In severe cases there is considerable swelling, some bleeding, no pain, and blistering. If untreated, gangrene may develop. Frostbite is treated by rapid rewarming.

Hemangioma (he-man′-jē-Ō-ma; *hem-* = blood; *-angi-* = blood vessel; *-oma* = tumor) Localized benign tumor of the skin and subcutaneous layer that results from an abnormal increase in blood vessels. One type is a **portwine stain,** a flat, pink, red, or purple lesion present at birth, usually at the nape of the neck.

Hives Reddened elevated patches of skin that are often itchy. Most commonly caused by infections, physical trauma, medications, emotional stress, food additives, and certain food allergies. Also called **urticaria** (ūr-ti-KAR-ē-a).

Keloid (KĒ-loid; *kelis* = tumor) An elevated, irregular darkened area of excess scar tissue caused by collagen formation during healing. It extends beyond the original injury and is tender and frequently painful. It occurs in the dermis and underlying subcutaneous tissue, usually after trauma, surgery, a burn, or severe acne; more common in people of African descent.

Keratosis (ker′-a-TŌ-sis; *kera-* = horn) Formation of a hardened growth of epidermal tissue, such as *solar keratosis*, a premalignant lesion of the sun-exposed skin of the face and hands.

Laceration (las-er-Ā-shun; *lacer-* = torn) An irregular tear of the skin.

Papule (PAP-ūl; *papula* = pimple) A small, round skin elevation less than 1 cm in diameter. One example is a pimple.

Pruritus (proo-RĪ-tus; *pruri-* = to itch) Itching, one of the most common dermatological disorders. It may be caused by skin disorders (infections), systemic disorders (cancer, kidney failure), psychogenic factors (emotional stress), or allergic reactions.

Topical In reference to a medication, applied to the skin surface rather than ingested or injected.

Wart Mass produced by uncontrolled growth of epithelial skin cells; caused by a papillomavirus. Most warts are noncancerous.

Chapter 4 | The Skeletal System

The skeletal system consists of the 206 named bones. There are 80 bones in the **axial skeleton** which includes the skull, hyoid, auditory ossicles, vertebral column and thorax. The **appendicular skeleton** has 126 bones and includes the pectoral girdles, upper limbs, pelvic girdle, and lower limbs. Bones provide support and protection for internal organs, as well aiding in body movements. Bones also store and release calcium, which is needed for proper function of most body tissues. Bone tissue produces red blood cells and stores minerals and triglycerides. The study of bone structure and the treatment of bone disorders is called **osteology**. Bones, muscles and joints together form an integrated system called the musculoskeletal system. The branch of medicine concerned with the prevention or correction of disorders of the musculoskeletal system is called **orthopedics.**

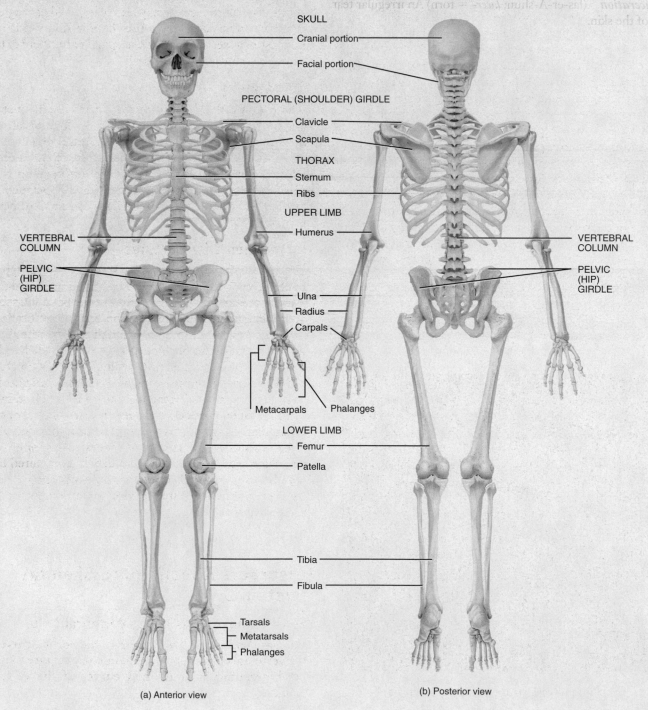

SKULL
Cranial portion
Facial portion

PECTORAL (SHOULDER) GIRDLE
Clavicle
Scapula

THORAX
Sternum
Ribs

UPPER LIMB
Humerus

VERTEBRAL COLUMN

PELVIC (HIP) GIRDLE

VERTEBRAL COLUMN

PELVIC (HIP) GIRDLE

Ulna
Radius
Carpals

Metacarpals   Phalanges

LOWER LIMB
Femur
Patella

Tibia
Fibula

Tarsals
Metatarsals
Phalanges

(a) Anterior view

(b) Posterior view

Tests, Procedures, and Techniques

■ Bone Grafting

Bone grafting generally consists of taking a piece of bone, along with its periosteum and nutrient artery, from one part of the body to replace missing bone in another part of the body. The transplanted bone restores the blood supply to the transplanted site, and healing occurs as in a fracture. The fibula is a common source of bone for grafting because even after a piece of the fibula has been removed, walking, running, and jumping can be normal. Recall that the tibia is the weight-bearing bone of the leg.

■ Bone Scan

A **bone scan** is a diagnostic procedure that takes advantage of the fact that bone is living tissue. A small amount of a radioactive tracer compound that is readily absorbed by bone is injected intravenously. The degree of uptake of the tracer is related to the amount of blood flow to the bone. A scanning device (gamma camera) measures the radiation emitted from the bones, and the information is translated into a photograph that can be read like an x-ray on a monitor. Normal bone tissue is identified by a consistent gray color throughout because of its uniform uptake of the radioactive tracer. Darker or lighter areas may indicate bone abnormalities. Darker areas called "hot spots" are areas of increased metabolism that absorb more of the radioactive tracer due to increased blood flow. Hot spots may indicate bone cancer, abnormal healing of fractures, or abnormal bone growth. Lighter areas called "cold spots" are areas of decreased metabolism that absorb less of the radioactive tracer due to decreased blood flow. Cold spots may indicate problems such as degenerative bone disease, decalcified bone, fractures, bone infections, Paget's disease, and rheumatoid arthritis. A bone scan

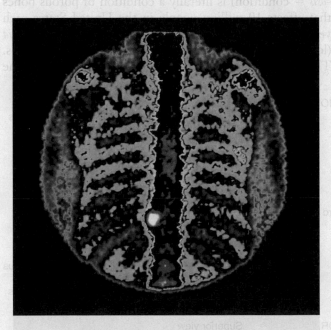

Hot spot (white area) in a cancerous vertebra

detects abnormalities 3 to 6 months sooner than standard x-ray procedures and exposes the patient to less radiation. A bone scan is the standard test for bone density screening, particularly important in screening for osteoporosis in females.

■ Caudal Anesthesia

Anesthetic agents that act on the sacral and coccygeal nerves are sometimes injected through the sacral hiatus, a procedure called **caudal anesthesia** or **epidural block.** The procedure is used most often to relieve pain during labor and to provide anesthesia to the perineal area. Because the sacral hiatus is between the sacral cornua, the cornua are important bony landmarks for locating the hiatus. Anesthetic agents may also be injected through the posterior sacral foramina. Since the injection site is inferior to the lowest portion of the spinal cord, there is little danger in damaging the cord.

■ Pelvimetry

Pelvimetry is the measurement of the size of the inlet and outlet of the birth canal, which may be done by ultrasonography or physical examination. Measurement of the pelvic cavity in pregnant females is important because the fetus must pass through the narrower opening of the pelvis at birth. A cesarean section is usually planned if it is determined that the pelvic cavity is too small to permit passage of the baby.

■ Treatments for Fractures

Treatments for fractures vary according to age, type of fracture, and the bone involved. The ultimate goals of fracture treatment are realignment of the bone fragments, immobilization to maintain realignment, and restoration of function. For bones to unite properly, the fractured ends must be brought into alignment, a process called **reduction,** commonly referred to as setting a fracture. In **closed reduction,** the fractured ends of a bone are brought into alignment by manual manipulation, and the skin remains intact. In **open reduction,** the fractured ends of a bone are brought into alignment by a surgical procedure in which internal fixation devices such as screws, plates, pins, rods, and wires are used. Following reduction, a fractured bone may be kept immobilized by a cast, sling, splint, elastic bandage, external fixation device, or a combination of these devices. ■

Disorders Affecting the Skeletal System

■ Abnormal Curves of the Vertebral Column

Various conditions may exaggerate the normal curves of the vertebral column, or the column may acquire a lateral bend, resulting in **abnor-mal curves** of the vertebral column.

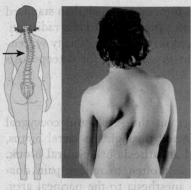

Scoliosis

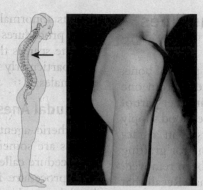

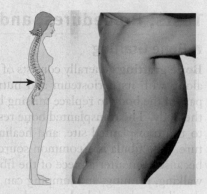

Kyphosis

Lordosis

Scoliosis (skō-lē-Ō-sis; *scolio* = crooked), the most common of the abnormal curves, is a lateral bending of the vertebral column, usually in the thoracic region. It may result from congenitally (present at birth) malformed vertebrae, chronic sciatica, paralysis of muscles on one side of the vertebral column, poor posture, or one leg being shorter than the other.

Signs of scoliosis include uneven shoulders and waist, one shoulder blade more prominent than the other, one hip higher than the other, and leaning to one side. In severe scoliosis (a curve greater than 70 degrees), breathing is more difficult and the pumping action of the heart is less efficient. Chronic back pain and arthritis of the vertebral column may also develop. Treatment options include wearing a back brace, physiotherapy, chiropractic care, and surgery (fusion of vertebrae and insertion of metal rods, hooks, and wires to reinforce the surgery).

Kyphosis (kī-FŌ-sis; *kyphoss-* = hump; *-osis* = condition) is an increase in the thoracic curve of the vertebral column (Figure 4.3b). In tuberculosis of the spine, vertebral bodies may partially collapse, causing an acute angular bending of the vertebral column. In the elderly, degeneration of the intervertebral discs leads to kyphosis. Kyphosis may also be caused by rickets and poor posture. It is also common in females with advanced osteoporosis. The term *round-shouldered* is an expression for mild kyphosis.

Lordosis (lor-DŌ-sis; *lord-* = bent backward), sometimes called *hollow back*, is an increase in the lumbar curve of the vertebral column. It may result from increased weight of the abdomen as in pregnancy or extreme obesity, poor posture, rickets, osteoporosis, or tuberculosis of the spine.

■ Herniated (Slipped) Disc

In their function as shock absorbers, intervertebral discs are constantly being compressed. If the anterior and posterior ligaments of the discs become injured or weakened, the pressure developed in the nucleus pulposus may be great enough to rupture the surrounding fibrocartilage (annulus fibrosus). If this occurs, the nucleus pulposus may herniate (protrude) posteriorly or into one of the adjacent vertebral bodies. This condition is

called a **herniated (slipped) disc.** Because the lumbar region bears much of the weight of the body, and is the region of the most flexing and bending, herniated discs most often occur in the lumbar area.

Frequently, the nucleus pulposus slips posteriorly toward the spinal cord and spinal nerves. This movement exerts pressure on the spinal nerves, causing local weakness and acute pain. If the roots of the sciatic nerve, which passes from the spinal cord to the foot, are compressed, the pain radiates down the posterior thigh, through the calf, and occasionally into the foot. If pressure is exerted on the spinal cord itself, some of its neurons may be destroyed. Treatment options include bed rest, medications for pain, physical therapy and exercises, and traction. A person with a herniated disc may also undergo a laminectomy, a procedure in which parts of the laminae of the vertebra and intervertebral disc are removed to relieve pressure on the nerves.

■ Osteoporosis

Osteoporosis (os'-tē-ō-pō-RŌ-sis; *por-* = passageway; *-osis* = condition) is literally a condition of porous bones that affects 10 million people in the United States each year. In addition, about 18 million people have *osteopenia* (low bone mass), which puts them at risk for osteoporosis. The basic problem is that bone resorption outpaces bone

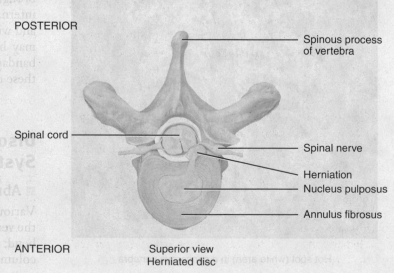

POSTERIOR

Spinous process of vertebra

Spinal cord

Spinal nerve

Herniation

Nucleus pulposus

Annulus fibrosus

ANTERIOR

Superior view
Herniated disc

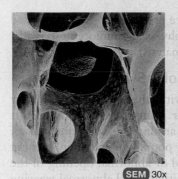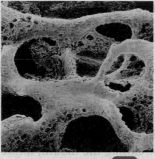

SEM 30x   SEM 30x

Normal bone   Osteoporotic bone

deposition. In large part this is due to depletion of calcium from the body—more calcium is lost in urine, feces, and sweat than is absorbed from the diet. Bone mass becomes so depleted that bones fracture, often spontaneously, under the mechanical stresses of everyday living. For example, a hip fracture might result from simply sitting down too quickly. In the United States, osteoporosis results in more than one and a half million fractures a year, mainly in the hip, wrist, and vertebrae. Osteoporosis afflicts the entire skeletal system. In addition to fractures, osteoporosis causes shrinkage of vertebrae, height loss, hunched backs, and bone pain.

The disorder primarily affects middle-aged and elderly people, 80% of them women. Older women suffer from osteoporosis more often than men for two reasons: (1) Women's bones are less massive than men's bones, and (2) production of estrogens in women declines dramatically at menopause, but production of the main androgen, testosterone, wanes gradually and only slightly in older men. Estrogens and testosterone stimulate osteoblast activity and synthesis of bone extracellular matrix. Besides gender, risk factors for developing osteoporosis include a family history of the disease, European or Asian ancestry, thin or small body build, an inactive lifestyle, cigarette smoking, a diet low in calcium and vitamin D, more than two alcoholic drinks a day, and the use of certain medications.

Osteoporosis is diagnosed by taking a family history and undergoing a *bone mineral density* (BMD) test. BMD tests are performed like x-rays and measure bone density. They can also be used to confirm a diagnosis of osteoporosis, determine the rate of bone loss, and monitor the effects of treatment.

Treatment options for osteoporosis are varied. With regard to nutrition, a diet high in calcium is important to reduce the risk of fractures. Vitamin D is necessary for the body to utilize calcium. In terms of exercise, regularly performing weight-bearing exercises has been shown to maintain and build bone mass. Among these exercises are walking, jogging, hiking, climbing stairs, playing tennis, and dancing. Resistance exercises, such as weight lifting, build bone strength and muscle mass.

Medications used to treat osteoporosis are generally of two types: (1) *Antiresorptive drugs* that slow down the progression of bone loss and (2) *bone-building drugs* that promote increasing bone mass. Among the antiresorptive drugs are (1) *bisphosphonates*, which inhibit osteoclasts (Fosamax®, Actonel®, Boniva®, and calcitonin); (2) selective estrogen receptor modulators, which mimic the effects of estrogens without unwanted side effects (Raloxifene®, Evista®); and (3) estrogen replacement therapy (ERT), which replaces estrogens lost during and after menopause (Premarin®), and hormone replacement therapy (HRT), which replaces estrogens and progesterone lost during and after menopause (Prempro®). HRT helps maintain and increase bone mass after menopause. Women on HRT have a slightly increased risk of stroke and blood clots. HRT also helps maintain and increase bone mass, and women on HRT have an increased risk of heart disease, breast cancer, stroke, blood clots, and dementia. Among the bone-forming drugs is parathyroid hormone (PTH), which stimulates osteoblasts to produce new bone (Fortes®). Others are under development.

Patellofemoral Stress Syndrome

Patellofemoral stress syndrome ("runner's knee") is one of the most common problems runners experience. During normal flexion and extension of the knee, the patella tracks (glides) superiorly and inferiorly in the groove between the femoral condyles. In patellofemoral stress syndrome, normal tracking does not occur; instead, the patella tracks laterally as well as superiorly and inferiorly, and the increased pressure on the joint causes aching or tenderness around or under the patella. The pain typically occurs after a person has been sitting for awhile, especially after exercise. It is worsened by squatting or walking down stairs. One cause of runner's knee is constantly walking, running, or jogging on the same side of the road. Because roads slope down on the sides, the knee that is closer to the center of the road endures greater mechanical stress because it does not fully extend during a stride. Other predisposing factors include running on hills, running long distances, and an anatomical deformity called knock-knee.

Rickets and Osteomalacia

Rickets and **osteomalacia** (os̄- tē-ō-ma-LĀ-shē-a; *malacia* = softness) are two forms of the same disease that result from inadequate calcification of the extracellular bone matrix, usually caused by a vitamin D deficiency. Rickets is a disease of children in which the growing bones become "soft" or rubbery and are easily deformed. Because new bone formed at the epiphyseal (growth) plates fails to ossify, bowed legs and deformities of the skull, rib cage, and pelvis are common. Osteomalacia is the adult counterpart of rickets, sometimes called *adult rickets*. New bone formed during remodeling fails to calcify, and the person experiences varying degrees of pain and tenderness in bones, especially the hip and legs. Bone fractures also result from minor trauma. Prevention and treatment for rickets and osteomalacia consists of the administration of adequate vitamin D.

■ Sinusitis

Sinusitis (sĭn-ū-SĪ-tis) is an inflammation of the mucous membrane of one or more paranasal sinuses. It may be caused by a microbial infection (virus, bacterium, or fungus), allergic reactions, nasal polyps, or a severely deviated nasal septum. If the inflammation or an obstruction blocks the drainage of mucus into the nasal cavity, fluid pressure builds up in the paranasal sinuses, and a sinus headache may develop. Other symptoms may include nasal congestion, inability to smell, fever, and cough. Treatment options include decongestant sprays or drops, oral decongestants, nasal corticosteroids, antibiotics, analgesics to relieve pain, warm compresses, and surgery.

■ Spina Bifida

Spina bifida (SPĪ-Ina BIF-i-da) is a congenital defect of the vertebral column in which laminae of L5 and/or S1 fail to develop normally and unite at the midline. The least serious form is called *spina bifida occulta*. It occurs in L5 or S1 and produces no symptoms. The only evidence of its presence is a small dimple with a tuft of hair in the overlying skin. Several types of spina bifida involve protrusion of meninges (membranes) and/or spinal cord through the defect in the laminae and are collectively termed *spina bifida cystica* because of the presence of a cyst-like sac protruding from the backbone. If the sac contains the meninges from the spinal cord and cerebrospinal fluid, the condition is called *spina bifida with meningocele* (me-NING-gō-sēl). If the spinal cord and/or its nerve roots are in the sac, the condition is called *spina bifida with meningomyelocele* (me-ning-gō-MĪ-ē-lō-sēl). The larger the cyst and the number of neural structures it contains, the more serious the neurological problems. In severe cases, there may be partial or complete paralysis, partial or complete loss of urinary bladder and bowel control, and the absence of reflexes. An increased risk of spina bifida is associated with low levels of a B vitamin called folic acid during pregnancy. Spina bifida may be diagnosed prenatally by a test of the

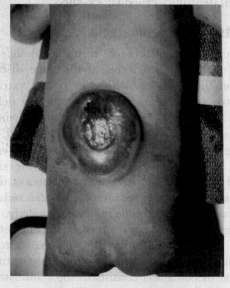

Cyst like sac of spina bifida

mother's blood for a substance produced by the fetus called alphafetoprotein, by sonography, or by amniocentesis (withdrawal of amniotic fluid for analysis).

■ Temporomandibular Joint Syndrome

One problem associated with the temporomandibular joint is **temporomandibular joint (TMJ) syndrome.** It is characterized by dull pain around the ear, tenderness of the jaw muscles, a clicking or popping noise when opening or closing the mouth, limited or abnormal opening of the mouth, headache, tooth sensitivity, and abnormal wearing of the teeth. TMJ syndrome can be caused by improperly aligned teeth, grinding or clenching the teeth, trauma to the head and neck, or arthritis. Treatments include application of moist heat or ice, limiting the diet to soft foods, administration of pain relievers such as aspirin, muscle retraining, use of a splint or bite plate to reduce clenching and teeth grinding (especially when worn at night), adjustment or reshaping of the teeth (orthodontic treatment), and surgery. ■

Additional Clinical Considerations

■ Black Eye

A **black eye** is a bruising around the eye, commonly due to an injury to the face, rather than an eye injury. In response to trauma, blood and other fluids accumulate in the space around the eye, causing the swelling and dark discoloration. One cause might be a blow to the sharp ridge just superior to the supraorbital margin that fractures the frontal bone, resulting in bleeding. Another is a blow to the nose. Certain surgical procedures (face lift, eyelid surgery, jaw surgery, or nasal surgery) can also result in black eyes.

■ Cleft Palate and Cleft Lip

Usually the palatine processes of the maxillary bones unite during weeks 10 to 12 of embryonic development. Failure to do so can result in one type of **cleft palate.** The condition may also involve incomplete fusion of the horizontal plates of the palatine bones. Another form of this condition, called **cleft lip,** involves a split in the upper lip. Cleft lip and cleft palate often occur together. Depending on the extent and position of the cleft, speech and swallowing may be affected. In addition, children with cleft palate tend to have many ear infections, which can lead to hearing loss. Facial and oral surgeons recommend closure of cleft lip during the first few weeks following birth, and surgical results are excellent. Repair of cleft palate typically is completed between 12 and 18 months of age, ideally before the child begins to talk. Because the palate is important for pronouncing consonants, speech therapy may be required, and orthodontic therapy may be needed to align the teeth. Again, results are usually excellent. Supplementation with folic acid (one of the B vitamins) during pregnancy decreases the incidence of cleft palate and cleft lip.

■ Deviated Nasal Septum

A **deviated nasal septum** is one that does not run along the midline of the nasal cavity. It deviates (bends) to one side. A blow to the nose can easily damage, or break, this delicate septum of bone and displace and damage the cartilage. Often, when a broken nasal septum heals, the bones and cartilage deviate to one side or the other. This deviated septum can block airflow into the constricted side of the nose, making it difficult to breathe through that half of the nasal cavity. The deviation usually occurs at the junction of the vomer bone with the septal cartilage. Septal deviations may also occur due to developmental abnormality. If the deviation is severe, it may block the nasal passageway entirely. Even a partial blockage may lead to infection. If inflammation occurs, it may cause nasal congestion, blockage of the paranasal sinus openings, chronic sinusitis, headache, and nosebleeds. The condition usually can be corrected or improved surgically.

■ Flatfoot and Clawfoot

The bones composing the arches of the foot are held in position by ligaments and tendons. If these ligaments and tendons are weakened, the height of the medial longitudinal arch may decrease or "fall." The result is **flatfoot**, the causes of which include excessive weight, postural abnormalities, weakened supporting tissues, and genetic predisposition. Fallen arches may lead to inflammation of the deep fascia of the sole (plantar fasciitis), Achilles tendinitis, shinsplints, stress fractures, bunions, and calluses. A custom-designed arch support often is prescribed to treat flatfoot.

Clawfoot is a condition in which the medial longitudinal arch is abnormally elevated. It is often caused by muscle deformities, such as may occur in diabetics whose neurological lesions lead to atrophy of muscles of the foot.

■ Fractures

The **clavicle** transmits mechanical force from the upper limb to the trunk. If the force transmitted to the clavicle is excessive, as when you fall on your outstretched arm, a **fractured clavicle** may result. A fractured clavicle may also result from a blow to the superior part of the anterior thorax. The clavicle is one of the most frequently broken bones in the body. Because the junction of the two curves of the clavicle is its weakest point, the clavicular midregion is the most frequent fracture site. Even in the absence of fracture, compression of the clavicle as a result of automobile accidents involving the use of shoulder harness seatbelts often causes damage to the median nerve, which lies between the clavicle and the second rib. A fractured clavicle is usually treated with a regular sling to keep the arm from moving outward.

Although any region of the hip girdle may fracture, the term **hip fracture** most commonly applies to a break in the bones associated with the hip joint—the head, neck, or trochanteric regions of the femur, or the bones that form the acetabulum. In the United States, 300,000 to 500,000 people sustain hip fractures each year. The incidence of hip fractures is increasing, due in part to longer life spans. Decreases in bone mass due to osteoporosis (which occurs more often in females), along with an increased tendency to fall, predispose elderly people to hip fractures.

Fractured clavicle

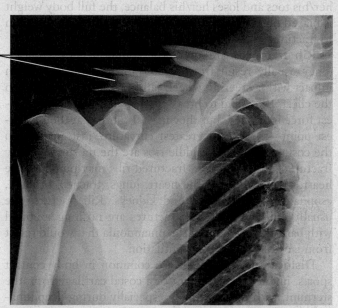

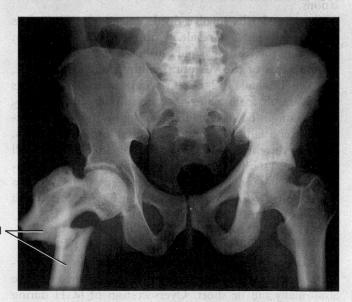
Fractured femur

Hip fractures often require surgical treatment, the goal of which is to repair and stabilize the fracture, increase mobility, and decrease pain. Sometimes the repair is accomplished by using surgical pins, screws, nails, and plates to secure the head of the femur. In severe hip fractures, the femoral head or the acetabulum of the hip bone may be replaced by prostheses (artificial devices). The procedure of replacing either the femoral head or the acetabulum is *hemiarthroplasty* (hem-ē-AR-thrō-plas-tē; *hemi-* = one half; *arthro-* = joint; *-plasty* = molding). Replacement of both the femoral head and acetabulum is *total hip arthroplasty*. The acetabular prosthesis is made of plastic, and the femoral prosthesis is metal; both are designed to withstand a high degree of stress. The prostheses are attached to healthy portions of bone with acrylic cement and screws.

Fractures of the metatarsals occur when a heavy object falls on the foot or when a heavy object rolls over the foot. Such fractures are also common among dancers, especially ballet dancers. If a ballet dancer is on the tip of her/his toes and loses her/his balance, the full body weight is placed on the metatarsals, causing one or more of them to fracture.

Rib fractures are the most common chest injuries. They usually result from direct blows, most often from impact with a steering wheel, falls, or crushing injuries to the chest. Ribs tend to break at the point where the greatest force is applied, but they may also break at their weakest point—the site of greatest curvature, just anterior to the costal angle. The middle ribs are the most commonly fractured. In some cases, fractured ribs may puncture the heart, great vessels of the heart, lungs, trachea, bronchi, esophagus, spleen, liver, and kidneys. Rib fractures are usually quite painful. Rib fractures are no longer bound with bandages because of the pneumonia that would result from lack of proper lung ventilation.

Dislocated ribs, which are common in body contact sports, involve displacement of a costal cartilage from the sternum, with resulting pain, especially during deep inhalations.

Separated ribs involve displacement of a rib and its costal cartilage; as a result, a rib may move superiorly, overriding the rib above and causing severe pain.

Fractures of the vertebral column often involve C1, C2, C4–T7, and T12–L2. Cervical or lumbar fractures usually result from a flexion-compression type of injury such as might be sustained in landing on the feet or buttocks after a fall or having a weight fall on the shoulders. Cervical vertebrae may be fractured or dislodged by a fall on the head with acute flexion of the neck, as might happen on diving into shallow water, or being thrown from a horse. Spinal cord or spinal nerve damage may occur as a result of fractures of the vertebral column.

■ Hormonal Abnormalities that Affect Height

Excessive or deficient secretion of hormones that normally control bone growth can cause a person to be abnormally tall or short. Oversecretion of hGH during childhood produces *giantism*, in which a person becomes much taller and heavier than normal. Undersecretion of hGH produces *pituitary dwarfism*, in which a person has short stature. (A dwarf has a normal-sized head and torso but small limbs; a midget has a proportioned head, torso, and limbs.) Because estrogens terminate growth at the epiphyseal (growth) plates, both men and women who lack estrogens or receptors for estrogens grow taller than

normal. Oversecretion of hGH during adulthood is called *acromegaly* (ak′-rō-MEG-a-lē). Although hGH cannot produce further lengthening of the long bones because the epiphyseal (growth) plates are already closed, the bones of the hands, feet, and jaws thicken and other tissues enlarge. In addition, the eyelids, lips, tongue, and nose enlarge, and the skin thickens and develops furrows, especially on the forehead and soles. hGH has been used to induce growth in short-statured children. ■

Medical Terminology

Chiropractic (kī-r ō-PRAK-tik; *cheir* = hand; *praktikos* = efficient). A holistic health care discipline that focuses on nerves, muscles, and bones. A *chiropractor* is a health care professional who is concerned with the diagnoses, treatment, and prevention of mechanical disorders of the musculoskeletal system and the effects of these disorders on the nervous system and health in general. Treatment involves using the hands to apply specific force to adjust joints of the body(manual adjustment), especially the vertebral column. Chiropractors may also use massage, heat therapy, ultrasound, electric stimulation, and acupuncture. Chiropractors often provide information about diet, exercise, changes in lifestyle, and stress management. Chiropractors do not prescribe drugs or perform surgery.

Clubfoot or talipes equinovarus (*-pes* = foot; *equino-* = horse) An inherited deformity in which the foot is twisted inferiorly and medially, and the angle of the arch is increased; occurs in 1 of every 1000 births. Treatment consists of manipulating the arch to a normal curvature by casts or adhesive tape, usually soon after birth. Corrective shoes or surgery may also be required.

Craniostenosis (krā-nē-ō-sten- ō -sis; *cranio-* = skull; *-stenosis*= narrowing). Premature closure of one or more cranial sutures during the first 18 to 20 months of life, resulting in a distorted skull. Premature closure of the sagittal suture produces a long narrow skull; premature closure of the coronal suture results in a broad skull. Premature closure of all sutures restricts brain growth and development; surgery is necessary to prevent brain damage.

Craniotomy (krā-nē-OT-ō-mē; *cranio-* = skull; -tome = cutting) Surgical procedure in which part of the cranium is removed. It may be performed to remove a blood clot, a brain tumor, or a sample of brain tissue for biopsy.

Genu valgum (JĒ-noo VAL-gum; *genu-* = knee; *valgum* = bent outward) A deformity in which the knees are abnormally close together and the space between the ankles is increased due to a lateral angulation of the tibia. Also called **knock-knee.**

Genu varum (JĒ-noo VAR-um; *varum* = bent toward the midline) A deformity in which the knees are abnormally separated and the lower limbs are bowed medially. Also called **bowleg.**

Hallux valgus (HAL-uks VAL-gus; *hallux* = great toe) Angulation of the great toe away from the midline of the body, typically caused by wearing tightly fitting shoes. When the great toe angles toward the next toe, there is a bony protrusion at the base of the great toe. Also called a **bunion.**

Laminectomy (lam′-i-NEK-tō -mē; *lamina-* = layer) Surgical procedure to remove a vertebral lamina. It may be performed to access the vertebral cavity and relieve the symptoms of a herniated disc.

Lumbar spine stenosis (*sten-* = narrowed) Narrowing of the spinal canal in the lumbar part of the vertebral column, due to hypertrophy of surrounding bone or soft tissues. It may be caused by arthritic changes in the intervertebral discs and is a common cause of back and leg pain.

Orthodontics (or-thō-DON-tiks) is the branch of dentistry concerned with the prevention and correction of poorly aligned teeth. The movement of teeth by braces places a stress on the bone that forms the sockets that anchor the teeth. In response to this artificial stress, osteoclasts and osteoblasts remodel the sockets so that the teeth align properly.

Osteoarthritis (os′-tē-ō-ar-THRĪ-tis; *arthr* = joint) The degeneration of articular cartilage such that the bony ends touch; the resulting friction of bone against bone worsens the condition. Usually associated with the elderly.

Osteogenic sarcoma (os′-tē-ō-JEN-ik sar-KŌ-ma; *sarcoma* = connective tissue tumor) Bone cancer that primarily affects osteoblasts and occurs most often in teenagers during their growth spurt; the most common sites are the metaphyses of the thigh bone (femur), shin bone (tibia), and arm bone (humerus). Metastases occur most often in lungs; treatment consists of multidrug chemotherapy and removal of the malignant growth, or amputation of the limb.

Osteomyelitis (os′-tē-ō-mī-e-LĪ-tis) An infection of bone characterized by high fever, sweating, chills, pain, nausea, pus formation, edema, and warmth over the affected bone and rigid overlying muscles. It is often caused by bacteria, usually *Staphylococcus aureus*. The bacteria may reach the bone from outside the body (through open fractures, penetrating wounds, or orthopedic surgical procedures); from other sites of infection in the body (abscessed teeth, burn infections, urinary tract infections, or upper respiratory infections) via the blood; and from adjacent soft tissue infections (as occurs in diabetes mellitus).

Osteopenia (os′-tēō-PĒ-nē-a; *penia* = poverty) Reduced bone mass due to a decrease in the rate of bone synthesis to a level too low to compensate for normal bone resorption; any decrease in bone mass below normal. An example is osteoporosis.

Spinal fusion (FŪ-zhun) Surgical procedure in which two or more vertebrae of the vertebral column are stabilized with a bone graft or synthetic device. It may be performed to treat a fracture of a vertebra or following removal of a herniated disc.

Whiplash injury Injury to the neck region due to severe hyperextension (backward tilting) of the head followed by severe hyperflexion (forward tilting) of the head, usually associated with a rear-end automobile collision. Symptoms are related to stretching and tearing of ligaments and muscles, vertebral fractures, and herniated vertebral discs.

A joint is the point of contact between two bones, between bone and cartilage, or between bone and teeth. Joints contribute to homeostasis by holding bones together in ways that allow for movement and flexibility. With **fibrous joints,** bones are held together by dense irregular connective tissue,

rich in collagen fibers. **Cartilaginous joints** hold bones together with by cartilage. **Synovial joints** have a synovial cavity, which is a space between the articulating bones. They are united by the dense irregular connective tissue of an articular capsule, and often by accessory ligaments.

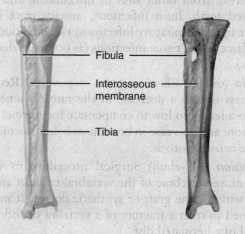

A fibrous joint

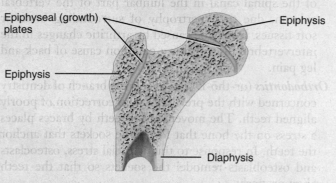

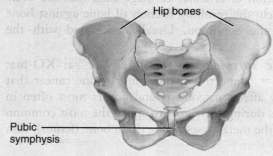

Cartilaginous joints

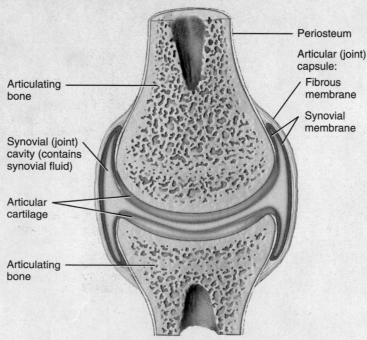

Typical synovial joint

Tests, Procedures, and Techniques

■ Arthroplasty

Joints that have been severely damaged by diseases such as arthritis, or by injury, may be replaced surgically with artificial joints in a procedure referred to as **arthroplasty** (AR-thrō-plas′-tē; *arthr-* = joint; *-plasty* = plastic repair of). Although most joints in the body can be repaired by arthroplasty, the ones most commonly replaced are the hips, knees, and shoulders. About 400,000 hip replacements and about 300,000 knee replacements are performed annually in the United States. During the procedure, the ends of the damaged bones are removed and metal, ceramic, or plastic components are fixed in place. The goals of arthroplasty are to relieve pain and increase range of motion.

Partial hip replacements involve only the femur. **Total hip replacements** involve both the acetabulum and head of the femur. The damaged portions of the acetabulum and the head of the femur are replaced by prefabricated prostheses (artificial devices). The acetabulum is shaped to accept the new socket, the head of the femur is removed, and the center of the femur is shaped to fit the femoral component. The acetabular component consists of a plastic such as polyethylene, and the femoral component is composed of a metal such as cobalt-chrome, titanium alloys, or stainless steel. These materials are designed to withstand a high degree of stress and to prevent a response by the immune system. Once the appropriate acetabular and femoral components are selected, they are attached to the healthy portion of bone with acrylic cement, which forms an interlocking mechanical bond.

Knee replacements are actually a resurfacing of cartilage and, like hip replacements, may be partial or total. In a **total knee replacement**, the damaged cartilage is removed from the distal end of the femur, proximal end of the tibia, and the back surface of the patella (if the back surface of the patella is not badly damaged, it may be left intact). The femur is reshaped and fitted with a metal femoral component and cemented in place. The tibia is reshaped and fitted with a plastic tibia component that is cemented in place. If the back surface of the patella is badly damaged, it is replaced with a plastic implant.

In a **partial knee replacement**, also called a **unicompartmental knee replacement**, only one side of the knee joint is replaced. Once the damaged cartilage is removed from the distal end of the femur, the femur is reshaped and a metal femoral component is cemented in place. Then the damaged cartilage from the proximal end of the tibia is removed, along with the meniscus. The tibia is reshaped and fitted with a plastic tibial component that is cemented into place.

Researchers are continually seeking to improve the strength of the cement and devise ways to stimulate bone growth around the implanted area. Potential complications of arthroplasty include infection, blood clots, loosening or dislocation of the replacement components, and nerve injury.

In a **partial knee replacement**, also called a **unicompartmental knee replacement**, only one side of the knee joint is replaced. Once the damaged cartilage is removed from the distal end of the femur, the femur is reshaped and a metal femoral component is cemented in place. Then the damaged cartilage from the proximal end of the tibia

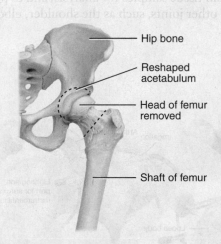

Preparation for total hip replacement

Hip bone

Reshaped acetabulum

Head of femur removed

Shaft of femur

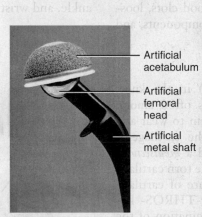

Components of an artificial hip joint

Artificial acetabulum

Artificial femoral head

Artificial metal shaft

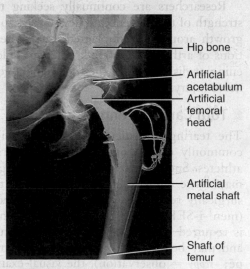

Radiograph of an artificial hip joint

Hip bone

Artificial acetabulum

Artificial femoral head

Artificial metal shaft

Shaft of femur

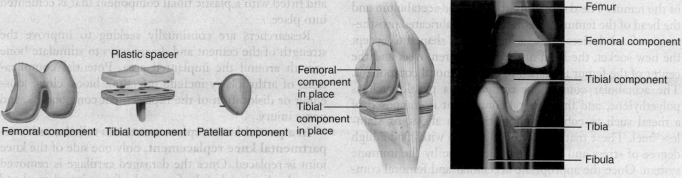

Preparation for total knee replacement

Plastic spacer

Femoral component   Tibial component   Patellar component

Components of artificial knee joint (isolated and in place)

Femoral component in place

Tibial component in place

Radiograph of a total knee replacement

is removed, along with the meniscus. The tibia is reshaped and fitted with a plastic tibial component that is cemented into place.

Researchers are continually seeking to improve the strength of the cement and devise ways to stimulate bone growth around the implanted area. Potential complications of arthroplasty include infection, blood clots, loosening or dislocation of the replacement components, and nerve injury.

■ Torn Cartilage and Arthroscopy

The tearing of articular discs (menisci) in the knee, commonly called **torn cartilage,** occurs often among athletes. Such damaged cartilage will begin to wear and may cause arthritis to develop unless the cartilage is surgically removed by a procedure called a *menisectomy* (men′-i-SEK-tō-mē). Surgical repair of the torn cartilage is required because of the avascular nature of cartilage and may be assisted by **arthroscopy** (ar-THROS- kō-pē; *-scopy* = observation), the visual examination of the interior of a joint, usually the knee, with an *arthroscope,* a lighted, pencil-thin instrument. Arthroscopy is used to determine the nature and extent of damage following knee injury and to monitor the progression of disease and

the effects of therapy. The insertion of surgical instruments through the arthroscope or other incisions enables a physician to remove torn cartilage and repair damaged cruciate ligaments in the knee; to remodel poorly formed cartilage; to obtain tissue samples for analysis; and to perform surgery on other joints, such as the shoulder, elbow, ankle, and wrist.

■

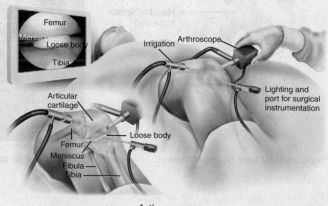

Arthroscopy

Disorders Affecting Joints

■ Ankylosing Spondylitis

Ankylosing spondylitis (ang′-ki-LŌ-sing spon′-di-LĪ-tis; *ankyle* = stiff; *spondyl* = vertebra) is an inflammatory disease of unknown origin that affects joints between vertebrae (intervertebral) and between the sacrum and hip bone (sacroiliac joint). The disease, which is more common in males, sets in between ages 20 and 40. It is characterized by pain and stiffness in the hips and lower back that progress upward along the backbone. Inflammation can lead to *ankylosis* (severe or complete loss of movement at a joint) and *kyphosis* (hunchback). Treatment consists of anti-inflammatory drugs, heat, massage, and supervised exercise.

■ Bursitis

An acute or chronic inflammation of a bursa, called **bursitis,** is usually caused by irritation from repeated, excessive exertion of a joint. The condition may also be caused by trauma, by an acute or chronic infection (including syphilis and tuberculosis), or by rheumatoid arthritis. Symptoms include pain, swelling, tenderness, and limited movement. Treatment may include oral anti-inflammatory agents and injections of cortisol-like steroids.

Gouty Arthritis

Uric acid (a substance that gives urine its name) is a waste product produced during the metabolism of nucleic acid (DNA and RNA) subunits. A person who suffers from **gout** (GOWT) either produces excessive amounts of uric acid or is not able to excrete as much as normal. The result is a buildup of uric acid in the blood. This excess acid then reacts with sodium to form a salt called sodium urate. Crystals of this salt accumulate in soft tissues such as the kidneys and in the cartilage of the ears and joints.

In **gouty arthritis,** sodium urate crystals are deposited in the soft tissues of the joints. Gout most often affects the joints of the feet, especially at the base of the big toe. The crystals irritate and erode the cartilage, causing inflammation, swelling, and acute pain. Eventually, the crystals destroy all joint tissues. If the disorder is untreated, the ends of the articulating bones fuse, and the joint becomes immovable. Treatment consists of pain relief (ibuprofen, naproxen, colchicine, and cortisone) followed by administration of allopurinol to keep uric acid levels low so that crystals do not form.

■ Lyme Disease

A spiral-shaped bacterium called *Borrelia burgdorferi* causes **Lyme disease,** named for the town of Lyme, Connecticut, where it was first reported in 1975. The bacteria are transmitted to humans mainly by deer ticks (*Ixodes dammini*). These ticks are so small that their bites often go unnoticed. Within a few weeks of the tick bite, a rash may appear at the site. Although the rash often resembles a bull's-eye target, there are many variations, and some people never develop a rash. Other symptoms include joint stiffness, fever and chills, headache, stiff neck, nausea, and low back pain. In advanced stages of the disease, arthritis is the main complication. It usually afflicts the larger joints such as the knee, ankle, hip, elbow, or wrist. Antibiotics are generally effective against Lyme disease, especially if they are given promptly. However, some symptoms may linger for years.

■ Osteoarthritis

Osteoarthritis (OA) (os′-tē-ō-ar-THRĪ-tis) is a degenerative joint disease in which joint cartilage is gradually lost. It results from a combination of aging, obesity, irritation of the joints, muscle weakness, and wear and abrasion. Commonly known as "wear-and-tear" arthritis, osteoarthritis is the most common type of arthritis.

Osteoarthritis is a progressive disorder of synovial joints, particularly weight-bearing joints. Articular cartilage deteriorates and new bone forms in the subchondral areas and at the margins of the joint. The cartilage slowly degenerates, and as the bone ends become exposed, spurs (small bumps) of new osseous tissue are deposited on them in a misguided effort by the body to protect against the friction. These spurs decrease the space of the joint cavity and restrict joint movement. Unlike rheumatoid arthritis (described next), osteoarthritis affects mainly the articular cartilage, although the synovial membrane often becomes inflamed late in the disease. Two major distinctions between osteoarthritis and rheumatoid arthritis are that osteoarthritis first afflicts the larger joints (knees, hips) and is due to wear and tear, whereas rheumatoid arthritis first strikes smaller joints and is an active attack of the cartilage. Osteoarthritis is the most common reason for hip- and knee-replacement surgery.

■ Rheumatoid Arthritis

Rheumatoid arthritis (RA) is an autoimmune disease in which the immune system of the body attacks its own tissues—in this case, its own cartilage and joint linings.

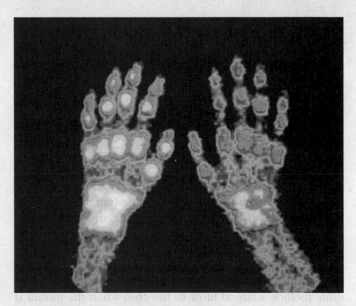

Gamma ray photograph of swollen joints (bright spots) due to RA

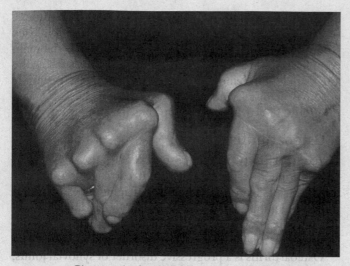

Photograph of an individual with severe RA

RA is characterized by inflammation of the joint, which causes swelling, pain, and loss of function. Usually, this form of arthritis occurs bilaterally: If one wrist is affected, the other is also likely to be affected, although often not to the same degree.

The primary symptom of RA is inflammation of the synovial membrane. If untreated, the membrane thickens, and synovial fluid accumulates. The resulting pressure causes pain and tenderness. The membrane then produces an abnormal granulation tissue, called pannus, that adheres to the surface of the articular cartilage and sometimes erodes the cartilage completely. When the cartilage is destroyed, fibrous tissue joins the exposed bone ends. The fibrous tissue ossifies and fuses the joint so that it becomes immovable—the ultimate crippling effect of RA. The growth of the granulation tissue causes the distortion of the fingers that characterizes hands of RA sufferers. ■

Additional Clinical Considerations

■ Dislocated Mandible

A **dislocation** (dis′-lō-KĀ-shun; *dis-* = apart) or *luxation* (luks-Ā-shun; *luxatio* = dislocation) is the displacement of a bone from a joint with tearing of ligaments, tendons, and articular capsules. It is usually caused by a blow or fall, although unusual physical effort may be a factor. For example, if the condylar processes of the mandible pass anterior to the articular tubercles when you yawn or take a large bite, a dislocated mandible (anterior displacement) may occur. When the mandible is displaced in this manner, the mouth remains wide open and the person is unable to close it. This may be corrected by pressing the thumbs downward on the lower molar teeth and pushing the mandible backward. Other causes of a dislocated mandible include a lateral blow to the chin when the mouth is open and a fracture of the mandible.

■ Knee Injuries

The knee joint is the joint most vulnerable to damage because it is a mobile, weight-bearing joint and its stability depends almost entirely on its associated ligaments and muscles. Further, there is no correspondence of the articulating bones. A **swollen knee** may occur immediately or hours after an injury. The initial swelling is due to escape of blood from damaged blood vessels adjacent to areas involving rupture of the anterior cruciate ligament, damage to synovial membranes, torn menisci, fractures, or collateral ligament sprains. Delayed swelling is due to excessive production of synovial fluid, a condition commonly referred to as "water on the knee." A common type of knee injury in football is **rupture of the tibial collateral ligaments,** often associated with tearing of the anterior cruciate ligament and medial meniscus (torn cartilage). Usually, a hard blow to the lateral side of the knee while the foot is fixed on the ground causes the damage. A **dislocated knee** refers to the displacement of the tibia relative to the femur. The most common type is dislocation anteriorly, resulting from hyperextension of the knee. A frequent consequence of a dislocated knee is damage to the popliteal artery.

■ Rotator Cuff Injury and Dislocated and Separated Shoulder

Rotator cuff injury is a strain or tear in the rotator cuff muscles and is a common injury among baseball pitchers, volleyball players, racket sports players, swimmers, and violinists, due to shoulder movements that involve vigorous circumduction. It also occurs as a result of wear and tear, aging, trauma, poor posture, improper lifting, and repetitive motions in certain jobs, such as placing items on a shelf above your head. Most often, there is tearing of the supraspinatus muscle tendon of the rotator cuff. This tendon is especially predisposed to wear-and-tear because of its location between the head of the humerus and acromion of the scapula, which compresses the tendon during shoulder movements. Poor posture and poor body mechanics also increase compression of the supraspinatus muscle tendon.

The joint most commonly dislocated in adults is the shoulder joint because its socket is quite shallow and the bones are held together by supporting muscles. Usually in a **dislocated shoulder,** the head of the humerus becomes displaced inferiorly, where the articular capsule is least protected. Dislocations of the mandible, elbow, fingers, knee, or hip are less common.

A **separated shoulder** refers to an injury of the acromioclavicular joint, a joint formed by the acromion of the scapula and the acromial end of the clavicle. This condition is usually the result of forceful trauma to the joint, as when the shoulder strikes the ground in a fall.

■ Sprain and Strain

A **sprain** is the forcible wrenching or twisting of a joint that stretches or tears its ligaments but does not dislocate the bones. It occurs when the ligaments are stressed beyond their normal capacity. Sprains also may damage surrounding blood vessels, muscles, tendons, or nerves. Severe sprains may be so painful that the joint cannot be moved. There is considerable swelling, which results from chemicals released by the damaged cells and hemorrhage of ruptured blood vessels. The lateral ankle joint is most often sprained; the lower back is another frequent location. A **strain** is a stretched or partially torn muscle or muscle and tendon. It often occurs when a muscle contracts suddenly and powerfully—such as the leg muscles of sprinters when they spring from the blocks.

■ Tennis Elbow, Little-League Elbow, and Dislocation of the Radial Head

Tennis elbow most commonly refers to pain at or near the lateral epicondyle of the humerus, usually caused by an improperly executed backhand. The extensor muscles strain or sprain, resulting in pain. **Little-league elbow** typically develops as a result of a heavy pitching schedule and/or a schedule that involves throwing curve balls, especially among youngsters. In this disorder, the elbow may enlarge, fragment, or separate.

A **dislocation of the radial head (nursemaid's elbow)** is the most common upper limb dislocation in children. In this injury, the head of the radius slides past or ruptures the radial annular ligament, a ligament that forms a collar around the head of the radius at the proximal radioulnar joint. Dislocation is most apt to occur when a strong pull is applied to the forearm while it is extended and supinated, for instance while swinging a child around with outstretched arms. ■

Medical Terminology

Arthralgia (ar-THRAL-jē-a; *arthr-* = joint; *-algia* = pain) Pain in a joint.

Arthritis is a form of rheumatism in which the joints are swollen, stiff, and painful. It afflicts about 45 million people in the United States, and is the leading cause of physical disability among adults over age 65.

Bursectomy (bur-SEK-tō-mē; *-ectomy* = removal of) Removal of a bursa.

Chondritis (kon-DRĪ-tis; *chondr-* = cartilage) Inflammation of cartilage.

Rheumatism (ROO-ma-tizm) is any painful disorder of the supporting structures of the body—bones, ligaments, tendons, or muscles—that is not caused by infection or injury.

Subluxation (sub-luks-Ā-shun) A partial or incomplete dislocation.

Synovitis (sin'-ō-VĪ-tis) Inflammation of a synovial membrane in a joint.

Chapter 6

The Muscular System

There are 700 individual muscles in your body that make up the muscular system. These include skeletal muscle tissue and connective tissue. The muscular system and mus-cle tissue contribute to homeostasis by stabilizing body position, producing movements, regulating organ volume, moving substances within the body, and producing heat.

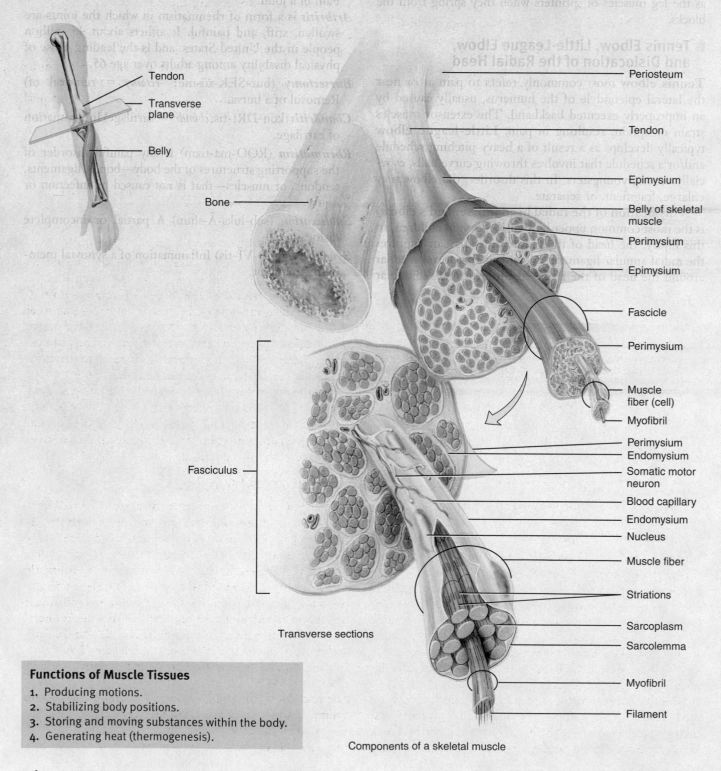

Transverse sections

Components of a skeletal muscle

Functions of Muscle Tissues
1. Producing motions.
2. Stabilizing body positions.
3. Storing and moving substances within the body.
4. Generating heat (thermogenesis).

Tests, Procedures, and Techniques

■ Electromyography

Electromyography (e-lek'-trō-mī-OG-ra-fē; *electro-* = electricity; *myo-* = muscle; *-graph* = to write) or **EMG** is a test that measures the electrical activity (muscle action potentials) in resting and contracting muscles. Normally, resting muscle produces no electrical activity; a slight contraction produces some electrical activity; and a more forceful contraction produces increased electrical activity. In the procedure, a ground electrode is placed over the muscle to be tested to eliminate background electrical activity. Then, a fine needle attached by wires to a recording instrument is inserted into the muscle. The electrical activity of the muscle is displayed as waves on an oscilloscope and heard through a loudspeaker.

EMG helps to determine if muscle weakness or paralysis is due to a malfunction of the muscle itself or the nerves supplying the muscle. EMG is also used to diagnose certain muscle disorders, such as muscular dystrophy.

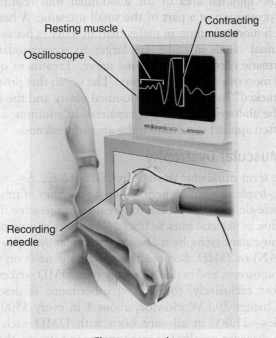

Resting muscle
Contracting muscle
Oscilloscope
Recording needle

Electromyography

■ Intramuscular Injections

An **intramuscular (IM) injection** penetrates the skin and subcutaneous tissue to enter the muscle itself. Intramuscular injections are preferred when prompt absorption is desired, when larger doses than can be given subcutaneously are indicated, or when the drug is too irritating to give subcutaneously. The common sites for intramuscular injections include the gluteus medius muscle of the buttock, lateral side of the thigh in the midportion of the vastus lateralis muscle, and the deltoid muscle of the shoulder.

Muscles in these areas, especially the gluteal muscles in the buttock, are fairly thick, and absorption is promoted by their extensive blood supply. To avoid injury, intramuscular injections are given deep within the muscle, away from

major nerves and blood vessels. Intramuscular injections have a faster speed of delivery than oral medications, but are slower than intravenous infusions.

■ Intubation During Anesthesia

When general anesthesia is administered during surgery, a total relaxation of the muscles results. Once the various types of drugs for anesthesia have been given (especially the paralytic agents), the patient's airway must be protected and the lungs ventilated because the muscles involved with respiration are among those paralyzed. Paralysis of the genioglossus muscle causes the tongue to fall posteriorly, which may obstruct the airway to the lungs. To avoid this, the mandible is either manually thrust forward and held in place (known as the "sniffing position"), or a tube is inserted from the lips through the laryngopharynx (inferior portion of the throat) into the trachea (endotracheal intubation). People can also be intubated nasally (through the nose).

Disorders Affecting the Muscular System

■ Bell's Palsy

Bell's palsy, also known as **facial paralysis,** is a unilateral paralysis of the muscles of facial expression. It is due to damage or disease of the facial (VII) nerve. Possible causes include inflammation of the facial nerve due to an ear infection, ear surgery that damages the facial nerve, or infection by the herpes simplex virus. The paralysis causes the entire side of the face to droop in severe cases. The person cannot wrinkle the forehead, close the eye, or pucker the lips on the affected side. Drooling and difficulty in swallowing also occur. Eighty percent of patients recover completely within a few weeks to a few months. For others, paralysis is permanent. The symptoms of Bell's palsy mimic those of a stroke.

■ Carpal Tunnel Syndrome

The **carpal tunnel** is a narrow passageway formed anteriorly by the flexor retinaculum and posteriorly by the carpal bones. Through this tunnel pass the median nerve, the most superficial structure, and the long flexor tendons for the digits (see Figure 11.18f). Structures within the carpal tunnel, especially the median nerve, are vulnerable to compression, and the resulting condition is called **carpal tunnel syndrome.** Compression of the median nerve leads to sensory changes over the lateral side of the hand and muscle weakness in the thenar eminence. This results in pain, numbness, and tingling of the fingers. The condition may be caused by inflammation of the digital tendon sheaths, fluid retention, excessive exercise, infection, trauma, and/or repetitive activities that involve flexion of the wrist, such as keyboarding, cutting hair, or playing the piano. Treatment may involve the use of nonsteroidal anti-inflammatory drugs (such as ibuprofen or aspirin), wearing a wrist splint, corticosteroid injections, or surgery to cut the flexor retinaculum and release presssure on the median nerve.

Compartment Syndrome

Skeletal muscles in the limbs are organized into functional units called *compartments*. In a disorder called **compartment syndrome,** some external or internal pressure constricts the structures within a compartment, resulting in damaged blood vessels and subsequent reduction of the blood supply (ischemia) to the structures within the compartment. Symptoms include pain, burning, pressure, pale skin, and paralysis. Common causes of compartment syndrome include crushing and penetrating injuries, contusion (damage to subcutaneous tissues without the skin being broken), muscle strain (overstretching of a muscle), or an improperly fitted cast. The pressure increase in the compartment can have serious consequences, such as hemorrhage, tissue injury, and edema (buildup of interstitial fluid). Because deep fasciae (connective tissue coverings) that enclose the compartments are very strong, accumulated blood and interstitial fluid cannot escape, and the increased pressure can literally choke off the blood flow and deprive nearby muscles and nerves of oxygen. One treatment option is **fasciotomy** (fash-ē-OT-ō-mē), a surgical procedure in which muscle fascia is cut to relieve the pressure. Without intervention, nerves can suffer damage, and muscles can develop scar tissue that results in permanent shortening of the muscles, a condition called *contracture.* If left untreated, tissues may die and the limb may no longer be able to function. Once the syndrome has reached this stage, amputation may be the only treatment option.

Fibromyalgia

Fibromyalgia (fī-brō-mī-AL-jē-a; *algia* = painful condition) is a painful, nonarticular rheumatic disorder that usually appears between the ages of 25 and 50. An estimated 3 million people in the United States suffer from fibromyalgia, which is 15 times more common in women than in men. The disorder affects the fibrous connective tissue components of muscles, tendons, and ligaments. A striking sign is pain that results from gentle pressure at specific "tender points." Even without pressure, there is pain, tenderness, and stiffness of muscles, tendons, and surrounding soft tissues. Besides muscle pain, those with fibromyalgia report severe fatigue, poor sleep, headaches, depression, and inability to carry out their daily activities. Treatment consists of stress reduction, regular exercise, application of heat, gentle massage, physical therapy, medication for pain, and a low-dose antidepressant to help improve sleep.

Impingement Syndrome

One of the most common causes of shoulder pain and dysfunction in athletes is known as **impingement syndrome,** which is sometimes confused with another common complaint, compartment syndrome. The repetitive movement of the arm over the head that is common in baseball, overhead racquet sports, lifting weights over the head, spiking a volleyball, and swimming puts these athletes at risk. Impingement syndrome may also be caused by a direct blow or stretch injury. Continual pinching of the supraspinatus tendon as a result of overhead motions causes it to become inflamed and results in pain. If movement is continued despite the pain, the tendon may degenerate near the attachment to the humerus and ultimately may tear away from the bone (rotator cuff injury). Treatment consists of resting the injured tendons, strengthening the shoulder through exercise, massage therapy, and surgery if the injury is particularly severe.

Inguinal Hernia

A **hernia** is a protrusion of an organ through a structure that normally contains it, which creates a lump that can be seen or felt through the skin's surface. The inguinal region is a weak area in the abdominal wall. It is often the site of an **inguinal hernia,** a rupture or separation of a portion of the inguinal area of the abdominal wall resulting in the protrusion of a part of the small intestine. A hernia is much more common in males than in females because the inguinal canals in males are larger to accommodate the spermatic cord and ilioinguinal nerve. Treatment of hernias most often involves surgery. The organ that protrudes is "tucked" back into the abdominal cavity and the defect in the abdominal muscles is repaired. In addition, a mesh is often applied to reinforce the area of weakness.

Muscular Dystrophy

The term **muscular dystrophy** (DIS-trō-fē; *dys-* = difficult;-*trophy* = nourishment) refers to a group of inherited muscle-destroying diseases that cause progressive degeneration of skeletal muscle fibers. The most common form of muscular dystrophy is *Duchenne muscular dystrophy* (doo-SHAN) or *DMD*. Because the mutated gene is on the X chromosome, and males have only one, DMD strikes boys almost exclusively. (Sex-linked inheritance is described in Chapter 29.) Worldwide, about 1 in every 3500 male babies—21,000 in all—are born with DMD each year. The disorder usually becomes apparent between the ages of 2 and 5, when parents notice the child falls often and has difficulty running, jumping, and hopping. By age 12 most boys with DMD are unable to walk. Respiratory or cardiac failure usually causes death by age 20.

In DMD, the gene that codes for the protein dystrophin is mutated, so little or no dystrophin is present in the sarcolemma. Without the reinforcing effect of dystrophin, the sarcolemma tears easily during muscle contraction, causing muscle fibers to rupture and die. The dystrophin gene was discovered in 1987, and by 1990 the first attempts were made to treat DMD patients with gene therapy. The muscles of three boys with DMD were injected with myoblasts bearing functional dystrophin genes, but only a few muscle fibers gained the ability to produce dystrophin. Similar clinical trials with additional patients have also failed. An alternative approach to the problem is to find a way to induce muscle fibers to produce the protein *utrophin*, which is similar to dystrophin. Experiments with dystrophin-deficient mice suggest this approach may work.

Myasthenia Gravis

Myasthenia gravis (mī-as-THĒ-nē-a GRAV-is; *mys-* = muscle; *aisthesis* = sensation) is an autoimmune disease that causes chronic, progressive damage of the neuromuscular junction. The immune system inappropriately produces antibodies that bind to and block some ACh receptors, thereby decreasing the number of functional ACh receptors at the motor end plates of skeletal muscles. Because 75% of patients with myasthenia gravis have hyperplasia or tumors of the thymus, it is thought that thymic abnormalities cause the disorder. As the disease progresses, more ACh receptors are lost. Thus, muscles become increasingly weaker, fatigue more easily, and may eventually cease to function.

Myasthenia gravis occurs in about 1 in 10,000 people and is more common in women, typically ages 20 to 40 at onset; men usually are ages 50 to 60 at onset. The muscles of the face and neck are most often affected. Initial symptoms include weakness of the eye muscles, which may produce double vision, and weakness of the throat muscles that may produce difficulty in swallowing. Later, the person has difficulty chewing and talking. Eventually the muscles of the limbs may become involved. Death may result from paralysis of the respiratory muscles, but often the disorder does not progress to this stage.

Anticholinesterase drugs such as pyridostigmine (Mestinon®) or neostigmine, the first line of treatment, act as inhibitors of acetylcholinesterase, the enzyme that breaks down ACh. Thus, the inhibitors raise the level of ACh that is available to bind with still-functional receptors. More recently, steroid drugs such as prednisone have been used with success to reduce antibody levels. Another treatment is plasmapheresis, a procedure that removes the antibodies from the blood. Often, surgical removal of the thymus (thymectomy) is helpful.

Plantar Fasciitis

Plantar fasciitis (fas-ē-Ī-tis) or **painful heel syndrome** is an inflammatory reaction due to chronic irritation of the plantar aponeurosis (fascia) at its origin on the calcaneus (heel bone). The aponeurosis becomes less elastic with age. This condition is also related to weight-bearing activities (walking, jogging, lifting heavy objects), improperly constructed or fitting shoes, excess weight (puts pressure on the feet), and poor biomechanics (flat feet, high arches, and abnormalities in gait may cause uneven distribution of weight on the feet). Plantar fasciitis is the most common cause of heel pain in runners and arises in response to the repeated impact of running. Treatments include ice, deep heat, stretching exercises, weight loss, prosthetics (such as shoe inserts or heel lifts), steroid injections, and surgery.

Strabismus

Strabismus (stra-BIZ-mus; *strabismos* = squinting) is a condition in which the two eyeballs are not properly aligned. This can be hereditary or it can be due to birth injuries, poor attachments of the muscles, problems with the brain's control center, or localized disease. Strabismus can be constant or intermittent. In strabismus, each eye sends an image to a different area of the brain and because the brain usually ignores the messages sent by one of the eyes, the ignored eye becomes weaker, hence "lazy eye" or *amblyopia*, develops. *External strabismus* results when a lesion in the oculomotor (III) nerve causes the eyeball to move laterally when at rest, and results in an inability to move the eyeball medially and inferiorly. A lesion in the abducens (VI) nerve results in *internal strabismus*, a condition in which the eyeball moves medially when at rest and cannot move laterally.

Treatment options for strabismus depend on the specific type of problem and include surgery, visual therapy (retraining the brain's control center), and orthoptics (eye muscle training to straighten the eyes).

Tenosynovitis

Tenosynovitis (ten′-ō-sin-ō-VĪ-tis) is an inflammation of the tendons, tendon sheaths, and synovial membranes surrounding certain joints. The tendons most often affected are at the wrists, shoulders, elbows (resulting in *tennis elbow*), finger joints (resulting in *trigger finger*), ankles, and feet. The affected sheaths sometimes become visibly swollen because of fluid accumulation. Tenderness and pain are frequently associated with movement of the body part. The condition often follows trauma, strain, or excessive exercise. Tenosynovitis of the dorsum of the foot may be caused by tying shoelaces too tightly. Gymnasts are prone to developing the condition as a result of chronic, repetitive, and maximum hyperextension at the wrists. Other repetitive movements involving activities such as typing, haircutting, carpentry, and assembly line work can also result in tenosynovitis.

Additional Clinical Considerations

Aerobic Training Versus Strength Training

Regular, repeated activities such as jogging or aerobic dancing increase the supply of oxygen-rich blood available to skeletal muscles for aerobic cellular respiration. By contrast, activities such as weight lifting rely more on anaerobic production of ATP through glycolysis. Such anaerobic activities stimulate synthesis of muscle proteins and result, over time, in increased muscle size (muscle hypertrophy). Athletes who engage in anaerobic training should have a diet that includes an adequate amount of proteins. This protein intake will allow the body to synthesize muscle proteins and to increase muscle mass. As a result, aerobic training builds endurance for prolonged activities; in contrast, anaerobic training builds muscle strength for short-term feats. **Interval training** is a workout regimen that incorporates both types of training—for example, alternating sprints with jogging.

Anabolic Steroids

The use of **anabolic steroids** by athletes has received widespread attention. These steroid hormones, similar to testosterone, are taken to increase muscle size and thus strength during athletic contests. However, the large doses needed to produce an effect have damaging, sometimes even devastating side effects, including liver cancer, kidney damage, increased risk of heart disease, stunted growth, wide mood swings, increased acne, and increased irritability and aggression. Additionally, females who take anabolic steroids may experience atrophy of the breasts and uterus, menstrual irregularities, sterility, facial hair growth, and deepening of the voice. Males may experience diminished testosterone secretion, atrophy of the testes, sterility, and baldness.

Back Injuries and Heavy Lifting

Next to headaches, medical experts note that back problems are the most common medical complaint that lead people to seek treatment. Second only to the common cold as the greatest cause of lost workdays, back injuries cost U.S. industry $10–14 billion in workers' compensation costs and about 100 million lost workdays annually.

The four factors associated with increased risk of back injury are amount of force, repetition, posture, and stress applied to the backbone. Poor physical condition, poor posture, lack of exercise, and excessive body weight contribute to the number and severity of sprains and strains. Back pain caused by a muscle strain or ligament sprain will normally heal within a short time and may never cause further problems. However, if ligaments and muscles are weak, discs in the lower back can become weakened and may herniate (rupture) with excessive lifting or a sudden fall. After years of back abuse, or with aging, the discs may simply wear out and cause chronic pain. Degeneration of the spine due to aging is often misdiagnosed as a sprain or strain.

Full flexion at the waist, as in touching your toes, overstretches the erector spinae muscles. Muscles that are overstretched cannot contract effectively. Straightening up from such a position is therefore initiated by the hamstring muscles on the back of the thigh and the gluteus maximus muscles of the buttocks. The erector spinae muscles join in as the degree of flexion decreases. Improperly lifting a heavy weight, however, can strain the erector spinae muscles. The result can be painful muscle spasms, tearing of tendons and ligaments of the lower back, and herniating of intervertebral discs. The lumbar muscles are adapted for maintaining posture, not for lifting. This is why it is important to bend at the knees and use the powerful extensor muscles of the thighs and buttocks while lifting a heavy load.

Creatine Supplementation

Creatine is both synthesized in the body (in the liver, kidneys, and pancreas) and derived from foods such as milk, red meat, and some fish. Adults need to synthesize and ingest a total of about 2 grams of creatine daily to make up for the urinary loss of creatinine, the breakdown product of creatine. Some studies have demonstrated improved performance during explosive movements, such as sprinting. Other studies, however, have failed to find a performance-enhancing effect of creatine supplementation. Moreover, ingesting extra creatine decreases the body's own synthesis of creatine, and it is not known whether natural synthesis recovers after long-term creatine supplementation. In addition, creatine supplementation can cause dehydration and may cause kidney dysfunction. Further research is needed to determine both the long-term safety and the value of creatine supplementation.

Exercise-Induced Muscle Damage

Comparison of electron micrographs of muscle tissue taken from athletes before and after intense exercise reveals considerable **exercise-induced muscle damage,** including torn sarcolemmas in some muscle fibers, damaged myofibrils, and disrupted Z discs. Microscopic muscle damage after exercise also is indicated by increases in blood levels of proteins, such as myoglobin and the enzyme creatine kinase, that are normally confined within muscle fibers. From 12 to 48 hours after a period of strenuous exercise, skeletal muscles often become sore. Such **delayed onset muscle soreness (DOMS)** is accompanied by stiffness, tenderness, and swelling. Although the causes of DOMS are not completely understood, microscopic muscle damage appears to be a major factor. In response to exercise-induced muscle damage, muscle fibers undergo repair: New regions of sarcolemma are formed to replace torn sarcolemmas, and more muscle proteins (including those of the myofibrils) are synthesized in the sarcoplasm of the muscle fibers.

Groin Pull

The five major muscles of the inner thigh function to move the legs medially. This muscle group is important in activities such as sprinting, hurdling, and horseback riding. A rupture or tear of one or more of these muscles can cause a **groin pull.** Groin pulls most often occur during sprinting or twisting, or from kicking a solid, perhaps stationary object. Symptoms of a groin pull may be sudden, or may not surface until the day after the injury, and include sharp pain in the inguinal region, swelling, bruising, or inability to contract the muscles. As with most strain injuries, treatment involves RICE therapy, which stands for *R*est, *I*ce, *C*ompression, and *E*levation. Ice should be applied immediately, and the injured part should be elevated and rested. An elastic bandage should be applied, if possible, to compress the injured tissue.

Hypotonia and Hypertonia

Hypotonia (*hypo-* = **below**) refers to decreased or lost muscle tone. Such muscles are said to be flaccid. Flaccid muscles are loose and appear flattened rather than rounded.

Certain disorders of the nervous system and disruptions in the balance of electrolytes (especially sodium, calcium, and, to a lesser extent, magnesium) may result in **flaccid paralysis,** which is characterized by loss of muscle tone, loss or reduction of tendon reflexes, and atrophy (wasting away) and degeneration of muscles.

Hypertonia (*hyper-* = above) refers to increased muscle tone and is expressed in two ways: spasticity or rigidity. **Spasticity** (spas-TIS-i-tē) is characterized by increased muscle tone (stiffness) associated with an increase in tendon reflexes and pathological reflexes (such as the Babinski sign, in which the great toe extends with or without fanning of the other toes in response to stroking the outer margin of the sole). Certain disorders of the nervous system and electrolyte disturbances such as those previously noted may result in **spastic paralysis,** partial paralysis in which the muscles exhibit spasticity. **Rigidity** refers to increased muscle tone in which reflexes are not affected, as occurs in tetanus.

■ Injury of Levator Ani and Urinary Stress Incontinence

During childbirth, the levator ani muscle supports the head of the fetus, and the muscle may be injured during a difficult childbirth or traumatized during an *episiotomy* (a cut made with surgical scissors to prevent or direct tearing of the perineum during the birth of a baby). The consequence of such injury may be **urinary stress incontinence,** that is, the leakage of urine whenever intra-abdominal pressure is increased—for example, during coughing. One way to treat urinary stress incontinence is to strengthen and tighten the muscles that support the pelvic viscera. This is accomplished by *Kegel exercises,* the alternate contraction and relaxation of muscles of the pelvic floor. To find the correct muscles, the person imagines that she is urinating and then contracts the muscles as if stopping in midstream. The muscles should be held for a count of three, then relaxed for a count of three. This should be done 5–10 times each hour—sitting, standing, and lying down. Kegel exercises are also encouraged during pregnancy to strengthen the muscles for delivery.

■ Muscular Atrophy and Hypertrophy

Muscular atrophy (A-trō-fē; *a-* = without, *-trophy* = nourishment) is a wasting away of muscles. Individual muscle fibers decrease in size because of progressive loss of myofibrils. Atrophy that occurs because muscles are not used is termed *disuse atrophy.* Bedridden individuals and people with casts experience disuse atrophy because the flow of nerve impulses (nerve action potentials) to inactive skeletal muscle is greatly reduced. The condition is reversible. If instead the nerve supply to a muscle is disrupted or cut, the muscle undergoes *denervation atrophy.* Over a period of 6 months to 2 years, the muscle shrinks to about one-fourth its original size, and the muscle fibers are irreversibly replaced by fibrous connective tissue.

Muscular hypertrophy is an increase in the diameter of muscle fibers due to increased production of myofibrils, mitochondria, sarcoplasmic reticulum, and other organelles. It results from very forceful, repetitive muscular activity, such as strength training. Because hypertrophied muscles contain more myofibrils, they are capable of more forceful contractions.

■ Pulled Hamstrings and Charley Horse

A strain or partial tear of the proximal hamstring muscles is referred to as **pulled hamstrings** or **hamstring strains.** Like pulled groins, they are common sports injuries in individuals who run very hard and/or are required to perform quick starts and stops. Sometimes the violent muscular exertion required to perform a feat tears away a part of the tendinous origins of the hamstrings, especially the biceps femoris, from the ischial tuberosity. This is usually accompanied by a contusion (bruising), tearing of some of the muscle fibers, and rupture of blood vessels, producing a hematoma (collection of blood) and sharp pain. Adequate training with good balance between the quadriceps femoris and hamstrings and stretching exercises before running or competing are important in preventing this injury.

The term **charley horse** (CHAR-lē- h-ōrs) is a slang term, and a popular name for a cramp or stiffness of muscles due to tearing of the muscle, followed by bleeding into the area. It is a common sports injury due to trauma or excessive activity and frequently occurs in the quadriceps femoris muscle, especially among football players.

■ Rigor Mortis

After death, cellular membranes become leaky. Calcium ions leak out of the sarcoplasmic reticulum into the cytosol and allow myosin heads to bind to actin. ATP synthesis ceases shortly after breathing stops, however, so the crossbridges cannot detach from actin. The resulting condition, in which muscles are in a state of rigidity (cannot contract or stretch), is called **rigor mortis** (rigidity of death). Rigor mortis begins 3–4 hours after death and lasts about 24 hours; then it disappears as proteolytic enzymes from lysosomes digest the crossbridges.

■ Running Injuries

Many individuals who jog or run sustain some type of running-related injury. Although such injuries may be minor, some can be quite serious. Untreated or inappropriately treated minor injuries may become chronic. Among runners, common sites of injury include the ankle, knee, calcaneal (Achilles) tendon, hip, groin, foot, and back. Of these, the knee often is the most severely injured area.

Running injuries are frequently related to faulty training techniques. This may involve improper or lack of sufficient warm-up routines, running too much, or running too soon after an injury. Or it might involve extended running on hard and/or uneven surfaces. Poorly constructed or worn-out running shoes can also contribute to injury, as can any biomechanical problem (such as a fallen arch) aggravated by running.

Most sports injuries should be treated initially with RICE therapy, which stands for *Rest*, *Ice*, *Compression*, and *Elevation*. Immediately apply ice, and rest and elevate the injured part. Then apply an elastic bandage, if possible, to compress the injured tissue. Continue using RICE for 2 to 3 days, and resist the temptation to apply heat, which may worsen the swelling. Follow-up treatment may include alternating moist heat and ice massage to enhance blood flow in the injured area. Sometimes it is helpful to take nonsteroidal anti-inflammatory drugs (NSAIDs) or to have local injections of corticosteroids. During the recovery period, it is important to keep active, using an alternative fitness program that does not worsen the original injury. This activity should be determined in consultation with a physician. Finally, careful exercise is needed to rehabilitate the injured area itself. Massage therapy may also be used to prevent or treat many sports injuries.

■ Shin Splint Syndrome

Shin splint syndrome, or simply **shin splints,** refers to pain or soreness along the tibia, specifically the medial, distal two-thirds. It may be caused by tendinitis of the anterior compartment muscles, especially the tibialis anterior muscle, inflammation of the periosteum (periostitis) around the tibia, or stress fractures of the tibia. The tendinitis usually occurs when poorly conditioned runners run on hard or banked surfaces with poorly supportive running shoes. The condition may also occur with vigorous activity of the legs following a period of relative inactivity or running in cold weather without proper warmup. The muscles in the anterior compartment (mainly the tibialis anterior) can be strengthened to balance the stronger posterior compartment muscles.

■ Stretching

The overall goal of **stretching** is to achieve normal range of motion of joints and mobility of soft tissues surrounding the joints. For most individuals, the best stretching routine involves *static stretching*, that is, slow sustained stretching that holds a muscle in a lengthened position. The muscles should be stretched to the point of slight discomfort (not pain) and held for about 30 seconds. Stretching should be done after warming up to increase the range of motion most effectively.

1. *Improved physical performance.* A flexible joint has the ability to move through a greater range of motion, which improves performance.
2. *Decreased risk of injury.* Stretching decreases resistance in various soft tissues so there is less likelihood of exceeding maximum tissue extensiblity during an activity (i.e., injuring the soft tissues).
3. *Reduced muscle soreness.* Stretching can reduce some of the muscle soreness that results after exercise.
4. *Improved posture.* Poor posture results from improper position of various parts of the body and the effects of gravity over a number of years. Stretching can help realign soft tissues to improve and maintain good posture. ■

Medical Terminology

Cramp A painful spasmodic contraction is known as a **cramp.** Cramps may be caused by inadequate blood flow to muscles, overuse of a muscle, dehydration, injury, holding a position for prolonged periods, and low blood levels of electrolytes, such as potassium.

Fasciculation (fa-sik-ū-LĀ-shun) An involuntary, brief twitch of an entire motor unit that is visible under the skin; it occurs irregularly and is not associated with movement of the affected muscle. Fasciculations may be seen in multiple sclerosis or in amyotrophic lateral sclerosis (Lou Gehrig's disease).

Fibrillation (fi-bri-LĀ-shun) A spontaneous contraction of a single muscle fiber that is not visible under the skin but can be recorded by electromyography. Fibrillations may signal destruction of motor neurons.

Muscle strain Tearing of a muscle because of forceful impact, accompanied by bleeding and severe pain. Also known as a *charley* horse or pulled muscle. It often occurs in contact sports and typically affects the quadriceps femoris muscle on the anterior surface of the thigh. The condition is treated by RICE therapy: rest (R), ice immediately after the injury (I), compression via a supportive wrap (C), and elevation of the limb (E).

Myalgia (mī-AL-jē-a;-algia = painful condition) Pain in or associated with muscles.

Myoma (mī-Ō-ma; -oma = tumor) A tumor consisting of muscle tissue.

Myomalacia (mī′-ō-ma-LĀ-shē-a;-malacia = soft) Pathological softening of muscle tissue.

Myositis (mī′-ō-SĪ-tis; -itis = inflammation of) Inflammation of muscle fibers (cells).

Myotonia (mī′-ō-TŌ-nē-a;-tonia = tension) Increased muscular excitability and contractility, with decreased power of relaxation; tonic spasm of the muscle.

Spasm A sudden involuntary contraction of a single muscle in a large group of muscles.

Tic A **tic** is a spasmodic twitching made involuntarily by muscles that are ordinarily under voluntary control. Twitching of the eyelid and facial muscles are examples of tics.

Tremor A **tremor** is a rhythmic, involuntary, purposeless contraction that produces a quivering or shaking movement.

Volkmann's contracture (FŌLK-manz kon-TRAK-tur; *contra-* = against) Permanent shortening (contracture) of a muscle due to replacement of destroyed muscle fibers by fibrous connective tissue, which lacks extensibility. Typically occurs in forearm flexor muscles. Destruction of muscle fibers may occur from interference with circulation caused by a tight bandage, a piece of elastic, or a cast.

Chapter 7 | The Nervous System

Together, the nervous system and the endocrine system share responsibility for maintaining homeostasis. The nervous system regulates body activities by responding rapidly using nerve impulses (action potentials) that provide communication with and regulation of most body tissues. The nervous system is also responsible for perception, behavior, and memories, and initiates all voluntary movements. The nervous system consists of two main subdivisions; the **central nervous system,** which includes the brain and spinal cord, and the **peripheral nervous system,** which includes all nervous tissue outside of the CNS. A **neurologist** is a physician who specializes in the diagnosis and treatment of disorders of the nervous system.

Tests, Procedures, and Techniques

■ Acupuncture

Acupuncture (ak-ū-PUNK-chur) is the use of fine needles (lasers, ultrasound, or electricity) inserted into specific exterior body locations (acupoints) and manipulated to relieve pain and provide therapy for various conditions. The placement of needles may cause the release of neurotransmitters such as endorphins, painkillers that may inhibit pain pathways.

■ Analgesia: Relief from Pain

Pain sensations sometimes occur out of proportion to minor damage, persist chronically due to an injury, or even appear

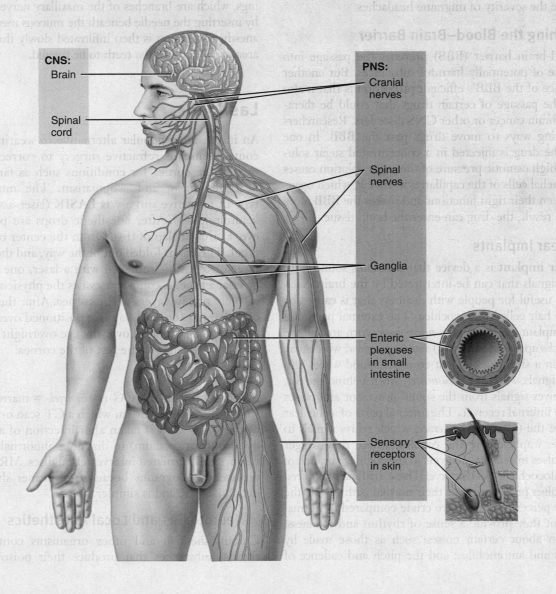

CNS:
- Brain
- Spinal cord

PNS:
- Cranial nerves
- Spinal nerves
- Ganglia
- Enteric plexuses in small intestine
- Sensory receptors in skin

for no obvious reason. In such cases, **analgesia** (*an-* = without; *-algesia* = pain) or pain relief is needed. Analgesic drugs such as aspirin and ibuprofen (for example, Advil® or Motrin®) block formation of prostaglandins, which stimulate nociceptors. Local anesthetics, such as Novocaine®, provide short-term pain relief by blocking conduction of nerve impulses along the axons of first-order pain neurons. Morphine and other opiate drugs alter the quality of pain perception in the brain; pain is still sensed, but it is no longer perceived as being so noxious. Many pain clinics use anticonvulsant and antidepressant medications to treat those suffering from chronic pain.

■ Biofeedback

Biofeedback is a technique in which an individual is provided with information regarding an autonomic response such as heart rate, blood pressure, or skin temperature. Various electronic monitoring devices provide visual or auditory signals about the autonomic responses. By concentrating on positive thoughts, individuals learn to alter autonomic responses. For example, biofeedback has been used to decrease heart rate and blood pressure and increase skin temperature in order to decrease the severity of migraine headaches.

■ Breaching the Blood–Brain Barrier

The blood-brain barrier (BBB) prevents the passage into brain tissue of potentially harmful substances. But another consequence of the BBB's efficient protection is that it also prevents the passage of certain drugs that could be therapeutic for brain cancer or other CNS disorders. Researchers are exploring ways to move drugs past the BBB. In one method, the drug is injected in a concentrated sugar solution. The high osmotic pressure of the sugar solution causes the endothelial cells of the capillaries to shrink, which opens gaps between their tight junctions and makes the BBB more leaky. As a result, the drug can enter the brain tissue.

■ Cochlear Implants

A **cochlear implant** is a device that translates sounds into electrical signals that can be interpreted by the brain. Such a device is useful for people with deafness that is caused by damage to hair cells in the cochlea. The external parts of a cochlear implant consist of (1) a *microphone* worn around the ear that picks up sound waves, (2) a *sound processor*, which may be placed in a shirt pocket, that converts sound waves into electrical signals, and (3) a *transmitter*, worn behind the ear, which receives signals from the sound processor and passes them to an internal receiver. The internal parts of a cochlear implant are the (1) *internal receiver*, which relays signals to (2) *electrodes* implanted in the cochlea, where they trigger nerve impulses in sensory neurons in the cochlear branch of the vestibulocochlear (VIII) nerve. These artificially induced nerve impulses propagate over their normal pathways to the brain. The perceived sounds are crude compared to normal hearing, but they provide a sense of rhythm and loudness; information about certain noises, such as those made by telephones and automobiles; and the pitch and cadence of speech. Some patients hear well enough with a cochlear implant to use the telephone.

■ Corneal transplant

Corneal transplant is a procedure in which a defective cornea is removed and a donor cornea of similar diameter is sewn in. It is the most common and most successful transplant operation. Since the cornea is avascular, antibodies in the blood that might cause rejection do not enter the transplanted tissue, and rejection rarely occurs. The shortage of donor corneas has been partially overcome by the development of artificial corneas made of plastic.

■ Dental Anesthesia

The inferior alveolar nerve, a branch of the mandibular nerve, supplies all the teeth in one-half of the mandible; it is often anesthetized in dental procedures. The same procedure will anesthetize the lower lip because the mental nerve is a branch of the inferior alveolar nerve. Because the lingual nerve runs very close to the inferior alveolar nerve near the mental foramen, it too is often anesthetized at the same time. For anesthesia to the upper teeth, the superior alveolar nerve endings, which are branches of the maxillary nerve, are blocked by inserting the needle beneath the mucous membrane. The anesthetic solution is then infiltrated slowly throughout the area of the roots of the teeth to be treated. ■

Lasik

An increasingly popular alternative to wearing glasses or contact lenses is refractive surgery to correct the curvature of the cornea for conditions such as farsightedness, nearsightedness, and astigmatism. The most common type of refractive surgery is **LASIK** (laser-assisted in-situ keratomileusis). After anesthetic drops are placed in the eye, a circular flap of tissue from the center of the cornea is cut. The flap is folded out of the way, and the underlying layer of cornea is reshaped with a laser, one microscopic layer at a time. A computer assists the physician in removing very precise layers of the cornea. After the sculpting is complete, the corneal flap is repositioned over the treated area. A patch is placed over the eye overnight and the flap quickly reattaches to the rest of the cornea.

■ Myelography

Myelography (mī-e-LOG-ra-fē; *myel-* = marrow, *-graph* = to write) is a procedure in which a CT scan or x-ray image of the spinal cord is taken after injection of a radiopaque dye (contrast medium) to diagnose abnormalities such as tumors and herniated intervertebral discs. MRI has largely replaced myelography because the former shows greater detail, is safer, and is simpler.

■ Neurotoxins and Local Anesthetics

Certain shellfish and other organisms contain **neurotoxins,** substances that produce their poisonous effects

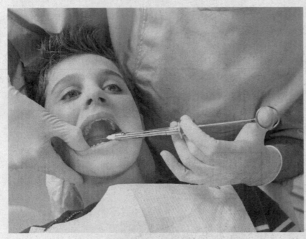

Local anesthetic-dental injection

by acting on the nervous system. One particularly lethal neurotoxin is tetrodotoxin (TTX), present in the viscera of Japanese pufferfish. TTX effectively blocks action potentials by inserting itself into voltage-gated Na^+ channels so they cannot open.

Local anesthetics are drugs that block pain and other somatic sensations. Examples include procaine (Novocaine®) and Lidocaine, which may be used to produce anesthesia in the skin during suturing of a gash, in the mouth during dental work, or in the lower body during childbirth. Like TTX, these drugs act by blocking the opening of voltage-gated Na^+ channels. Action potentials cannot propagate past the obstructed region, so pain signals do not reach the CNS.

Localized cooling of a nerve can also produce an anesthetic effect because axons propagate action potentials at lower speeds when cooled. The application of ice to injured tissue can reduce pain because propagation of the pain sensations along axons is partially blocked.

■ Reflexes and Diagnosis

Reflexes are often used for diagnosing disorders of the nervous system and locating injured tissue. If a reflex ceases to function or functions abnormally, the physician may suspect that the damage lies somewhere along a particular conduction pathway. Many somatic reflexes can be tested simply by tapping or stroking the body. Among the somatic reflexes of clinical significance are the following:

- **Patellar reflex (knee jerk).** This stretch reflex involves extension of the leg at the knee joint by contraction of the quadriceps femoris muscle in response to tapping the patellar ligament. This reflex is blocked by damage to the sensory or motor nerves supplying the muscle or to the integrating centers in the second, third, or fourth lumbar segments of the spinal cord. It is often absent in people with chronic diabetes mellitus or neurosyphilis, both of which cause degeneration of nerves. It is exaggerated in disease or injury involving certain motor tracts descending from the higher centers of the brain to the spinal cord.

- **Achilles reflex (ankle jerk).** This stretch reflex involves extension (plantar flexion) of the foot by contraction of the gastrocnemius and soleus muscles in response to tapping the calcaneal (Achilles) tendon. Absence of the Achilles reflex indicates damage to the nerves supplying the posterior leg muscles or to neurons in the lumbosacral region of the spinal cord. This reflex may also disappear in people with chronic diabetes, neurosyphilis, alcoholism, and subarachnoid hemorrhages. An exaggerated Achilles reflex indicates cervical cord compression or a lesion of the motor tracts of the first or second sacral segments of the cord.

- **Babinski sign.** This reflex results from gentle stroking of the lateral outer margin of the sole. The great toe dorsiflexes, with or without a lateral fanning of the other toes. This phenomenon normally occurs in children under 1½ years of age and is due to incomplete myelination of fibers in the corticospinal tract. A positive Babinski sign after age 1½ is abnormal and indicates an interruption of the corticospinal tract as the result of a lesion of the tract, usually in the upper portion. The normal response after age 1½ is the **plantar flexion reflex,** or **negative Babinski**—a curling under of all the toes.

- **Abdominal reflex.** This reflex involves contraction of the muscles that compress the abdominal wall in response to stroking the side of the abdomen. The response is an abdominal muscle contraction that causes the umbilicus to move in the direction of the stimulus. Absence of this reflex is associated with lesions of the corticospinal tracts. It may also be absent because of lesions of the peripheral nerves, lesions of integrating centers in the thoracic part of the cord, or multiple sclerosis.

Most autonomic reflexes are not practical diagnostic tools because it is difficult to stimulate visceral effectors, which are deep inside the body. An exception is the pupillary light reflex, in which the pupils of both eyes decrease in diameter when either eye is exposed to light. Because the reflex arc includes synapses in lower parts of the brain, the **absence of a normal pupillary light reflex** may indicate brain damage or injury.

■ Spinal Tap

In a **spinal tap (lumbar puncture),** a local anesthetic is given, and a long needle is inserted into the subarachnoid space. During this procedure, the patient lies on the side with the vertebral column flexed as in assuming the fetal position. Flexion of the vertebral column increases the distance between the spinous processes of the vertebrae, which allows easy access to the subarachnoid space. As you will soon learn, the spinal cord ends around the second lumbar vertebra (L2); however, the spinal meninges extend to the second sacral vertebra (S2). Between vertebrae L2 and S2 the spinal meninges are present, but the spinal cord is absent. Consequently, a spinal tap is normally performed in adults between vertebrae L3 and L4 or L4 and L5 because this region provides safe access to the subarachnoid space without the risk of damaging the spinal cord. (A line

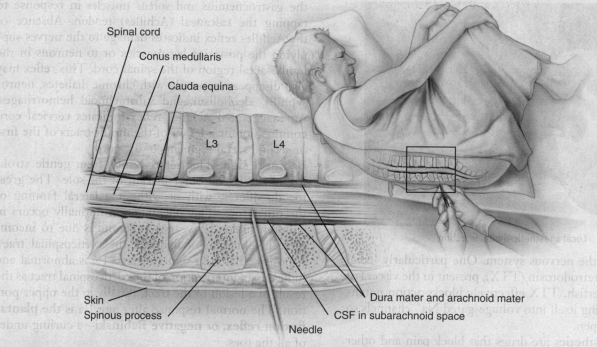

Spinal cord
Conus medullaris
Cauda equina
L3
L4
Skin
Spinous process
Needle
Dura mater and arachnoid mater
CSF in subarachnoid space

Spinal tap

drawn across the highest points of the iliac crests, called the *supracristal line*, passes through the spinous process of the fourth lumbar vertebra.) A spinal tap is used to withdraw cerebrospinal fluid (CSF) for diagnostic purposes; to introduce antibiotics, contrast media for myelography, or anesthetics; to administer chemotherapy; to measure CSF pressure; and/or to evaluate the effects of treatment for diseases such as meningitis.

Disorders Affecting the Nervous System

■ Age-related Macular Disease

Age-related macular disease (AMD), also known as *macular degeneration*, is a degenerative disorder of the retina in persons 50 years of age and older. In AMD, abnormalities occur in the region of the macula lutea, which is ordinarily the area of most acute vision. Victims of advanced AMD retain their peripheral vision but lose the ability to see straight ahead. For instance, they cannot see facial features to identify a person in front of them. AMD is the leading cause of blindness in those over age 75, afflicting 13 million Americans, and is 2.5 times more common in pack-a-day smokers than in nonsmokers. Initially, a person may experience blurring and distortion at the center of the visual field. In "dry" AMD, central vision gradually diminishes because the pigmented layer atrophies and degenerates. There is no effective treatment. In about 10% of cases, dry AMD progresses to "wet" AMD, in which new blood vessels form in the choroid and leak plasma or blood under the retina. Vision loss can be slowed by using laser surgery to destroy the leaking blood vessels.

■ Alzheimer Disease

Alzheimer disease (ALTZ-hī-mer) or **AD** is a disabling senile dementia, the loss of reasoning and ability to care for oneself, that afflicts about 11% of the population over age 65. In the United States, about 4 million people suffer from AD. Claiming over 100,000 lives a year, AD is the fourth leading cause of death among the elderly, after heart disease, cancer, and stroke. The cause of most AD cases is still unknown, but evidence suggests it is due to a combination of genetic factors, environmental or lifestyle factors, and the aging process. Mutations in three different genes (coding for presenilin-1, presenilin-2, and amyloid precursor protein) lead to early-onset forms of AD in afflicted families but account for less than 1% of all cases. An environmental risk factor for developing AD is a history of head injury. A similar dementia occurs in boxers, probably caused by repeated blows to the head.

Individuals with AD initially have trouble remembering recent events. They then become confused and forgetful, often repeating questions or getting lost while traveling to familiar places. Disorientation grows, memories of past events disappear, and episodes of paranoia, hallucination, or violent changes in mood may occur. As their minds continue to deteriorate, they lose their ability to read, write, talk, eat, or walk. The disease culminates in dementia. A person with AD usually dies of some complication that afflicts bedridden patients, such as pneumonia.

At autopsy, brains of AD victims show three distinct structural abnormalities:

1. *Loss of neurons that liberate acetylcholine.* A major center of neurons that liberate ACh is the nucleus basalis, which is below the globus pallidus. Axons of these neurons project widely throughout the cerebral cortex

and limbic system. Their destruction is a hallmark of Alzheimer disease.

2. *Beta-amyloid plaques,* clusters of abnormal proteins deposited outside neurons.

3. *Neurofibrillary tangles,* abnormal bundles of filaments inside neurons in affected brain regions. These filaments consist of a protein called *tau* that has been *hyperphosphorylated* (too many phosphate groups have been added to it).

Drugs that inhibit acetylcholinesterase (AChE), the enzyme that inactivates ACh, improve alertness and behavior in about 5% of AD patients. Tacrine®, the first anticholinesterase inhibitor approved for treatment of AD in the United States, has significant side effects and requires dosing four times a day. Donepezil®, approved in 1998, is less toxic to the liver and has the advantage of once-a-day dosing. Some evidence suggests that vitamin E (an antioxidant), estrogen, ibuprofen, and ginkgo biloba extract may have slight beneficial effects in AD patients. In addition, researchers are currently exploring ways to develop drugs that will prevent beta-amyloid plaque formation by inhibiting the enzymes involved in beta-amyloid synthesis and by increasing the activity of the enzymes involved in beta-amyloid degradation. Researchers are also trying to develop drugs that will reduce the formation of neurofibrillary tangles by inhibiting the enzymes that hyperphosphorylate tau.

■ Amnesia

Amnesia (am-NĒ-zē-a = forgetfulness) refers to the lack or loss of memory. It is a total or partial inability to remember past experiences. In *anterograde amnesia*, there is memory loss for events that occur *after* the trauma or disease that caused the condition. In other words, it is an inability to form new memories. In *retrograde amnesia*, there is a memory loss for events that occurred *before* the trauma or disease that caused the condition. In other words, it is an inability to recall past events.

■ Amyotrophic Lateral Sclerosis

Amyotrophic lateral sclerosis (ALS) (ā-mī-ō-TROF-ik; *a-* = without; *myo-* = muscle; *trophic* = nourishment) is a progressive degenerative disease that attacks motor areas of the cerebral cortex, axons of upper motor neurons in the lateral white columns (corticospinal and rubrospinal tracts), and lower motor neuron cell bodies. It causes progressive muscle weakness and atrophy. ALS often begins in sections of the spinal cord that serve the hands and arms but rapidly spreads to involve the whole body and face, without affecting intellect or sensations. Death typically occurs in 2 to 5 years. ALS is commonly known as *Lou Gehrig's disease* after the New York Yankees baseball player who died from it at age 37 in 1941.

Inherited mutations account for about 15% of all cases of ALS (familial ALS). Noninherited (sporadic) cases of ALS appear to have several implicating factors. According

to one theory there is a buildup in the synaptic cleft of the neurotransmitter glutamate released by motor neurons due to a mutation of the protein that normally deactivates and recycles the neurotransmitter. The excess glutamate causes motor neurons to malfunction and eventually die. The drug riluzole, which is used to treat ALS, reduces damage to motor neurons by decreasing the release of glutamate. Other factors may include damage to motor neurons by free radicals, autoimmune responses, viral infections, deficiency of nerve growth factor, apoptosis (programmed cell death), environmental toxins, and trauma.

In addition to riluzole, ALS is treated with drugs that relieve symptoms such as fatigue, muscle pain and spasticity, excessive saliva, and difficulty sleeping. The only other treatment is supportive care provided by physical, occupational, and speech therapists; nutritionists; social workers; and home care and hospice nurses.

■ Attention Deficit Hyperactivity Disorder

Attention deficit hyperactivity disorder (ADHD) is a learning disorder characterized by poor or short attention span, a consistent level of hyperactivity, and a level of impulsiveness inappropriate for the child's age. ADHD is believed to affect about 5% of children and is diagnosed 10 times more in boys than girls. The condition typically begins in childhood and continues into adolescence and adulthood. Symptoms of ADHD develop in early childhood, often before age four, and include difficulty in organizing and finishing tasks, lack of attention to details, short attention span and inability to concentrate, difficulty following instructions, talking excessively and frequently interrupting others, frequent running or excessive climbing, inability to play quietly alone, and difficulty waiting or taking turns.

The causes of ADHD are not fully understood, but it does have a strong genetic component. Some evidence also suggests that ADHD is related to problems with neurotransmitters. In addition, recent imaging studies have demonstrated that people with ADHD have less nervous tissue in specific regions of the brain such as the frontal and temporal lobes, caudate nucleus, and cerebellum. Treatment may involve remedial education, behavioral modification techniques, restructuring routines, and drugs that calm the child and help focus attention.

■ Disorders of the Basal Ganglia

Disorders of the basal ganglia can affect body movements, cognition, and behavior. Uncontrollable shaking (tremor) and muscle rigidity (stiffness) are hallmark signs of Parkinson disease (PD). In this disorder, dopamine-releasing neurons that extend from the substantia nigra to the putamen and caudate nucleus degenerate.

Huntington disease (HD) is an inherited disorder in which the caudate nucleus and putamen degenerate, with loss of neurons that normally release GABA or acetylcholine. A key sign of HD is **chorea** (KŌ-rē-a = a dance), in which rapid, jerky movements occur involuntarily and without purpose. Progressive mental deterioration also occurs.

Symptoms of HD often do not appear until age 30 or 40. Death occurs 10 to 20 years after symptoms first appear.

Tourette syndrome is a disorder that is characterized by involuntary body movements (motor tics) and the use of inappropriate or unnecessary sounds or words (vocal tics). Although the cause is unknown, research suggests that this disorder involves a dysfunction of the cognitive neural circuits between the basal ganglia and the prefrontal cortex.

Some psychiatric disorders, such as schizophrenia and obsessive-compulsive disorder, are thought to involve dysfunction of the behavioral neural circuits between the basal ganglia and the limbic system. In **schizophrenia,** excess dopamine activity in the brain causes a person to experience delusions, distortions of reality, paranoia, and hallucinations. People who have **obsessive-compulsive disorder (OCD)** experience repetitive thoughts (obsessions) that cause repetitive behaviors (compulsions) that they feel obligated to perform. For example, a person with OCD might have repetitive thoughts about someone breaking into the house; these thoughts might drive that person to check the doors of the house over and over again (for minutes or hours at a time) to make sure that they are locked.

Cerebral palsy

Cerebral palsy (CP) is a motor disorder that results in the loss of muscle control and coordination; caused by damage of the motor areas of the brain during fetal life, birth, or infancy. Radiation during fetal life, temporary lack of oxygen during birth, and hydrocephalus during infancy may also cause cerebral palsy.

Cerebrovascular Accident

The most common brain disorder is a **cerebrovascular accident (CVA),** also called a **stroke** or **brain attack.** CVAs affect 500,000 people each year in the United States and represent the third leading cause of death, behind heart attacks and cancer. A CVA is characterized by abrupt onset of persisting neurological symptoms, such as paralysis or loss of sensation, that arise from destruction of brain tissue. Common causes of CVAs are intracerebral hemorrhage (from a blood vessel in the pia mater or brain), emboli (blood clots), and atherosclerosis (formation of cholesterol-containing plaques that block blood flow) of the cerebral arteries.

Among the risk factors implicated in CVAs are high blood pressure, high blood cholesterol, heart disease, narrowed carotid arteries, transient ischemic attacks (TIAs; discussed next), diabetes, smoking, obesity, and excessive alcohol intake.

A clot-dissolving drug called tissue plasminogen activator (t-PA) is now being used to open up blocked blood vessels in the brain. The drug is most effective when administered *within three hours* of the onset of the CVA, however, and is helpful only for CVAs due to a blood clot. Use of t-PA can decrease the permanent disability associated with these types of CVAs by 50%. New studies show that "cold therapy" might be successful in limiting the amount of residual damage from a CVA. These "cooling" therapies developed from knowledge obtained following examination of cold water drowning victims. States of hypothermia seem to trigger a survival response in which the body requires less oxygen. Some commercial companies now provide "CVA survival kits," which include cooling blankets that can be kept in the home.

Depression

Depression is a disorder that affects over 18 million people each year in the United States. People who are depressed feel sad and helpless, have a lack of interest in activities that they once enjoyed, and experience suicidal thoughts. There are several types of depression. A person with **major depression** experiences symptoms of depression that last for more than two weeks. A person with **dysthymia** (dis-THĪ-mē-a) experiences episodes of depression that alternate with periods of feeling normal. A person with **bipolar disorder,** or *manic-depressive illness,* experiences recurrent episodes of depression and extreme elation (mania). A person with **seasonal affective disorder (SAD)** experiences depression during the winter months, when daylength is short. Although the exact cause of depression is unknown, research suggests that depression is linked to an imbalance of the neurotransmitters serotonin, norepinephrine, and dopamine in the brain. Factors that may contribute to depression include heredity, stress, chronic illnesses, certain personality traits (such as low self-esteem), and hormonal changes.

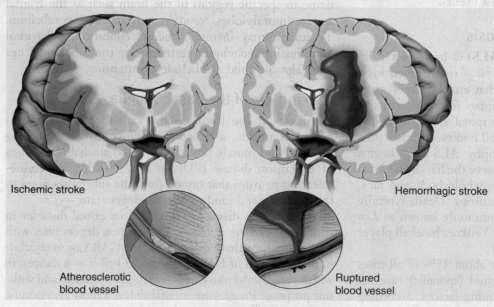

Ischemic stroke

Hemorrhagic stroke

Atherosclerotic blood vessel

Ruptured blood vessel

CV accident

Medication is the most common treatment for depression. For example, **selective serotonin reuptake inhibitors (SSRIs)** are drugs that provide relief from some forms of depression. By inhibiting reuptake of serotonin by serotonin transporters, SSRIs prolong the activity of this neurotransmitter at synapses in the brain. SSRIs include fluoxetine (Prozac®), paroxetine (Paxil®), and sertraline (Zoloft®).

Encephalitis

Encephalitis (en′-sef-a-LĪ-tis) is an acute inflammation of the brain caused by either a direct attack by any of several viruses or an allergic reaction to any of the many viruses that are normally harmless to the central nervous system. If the virus affects the spinal cord as well, the condition is called **encephalomyelitis.**

Epilepsy

Epilepsy is characterized by short, recurrent attacks of motor, sensory, or psychological malfunction, although it almost never affects intelligence. The attacks, called *epileptic seizures,* afflict about 1% of the world's population. They are initiated by abnormal, synchronous electrical discharges from millions of neurons in the brain, perhaps resulting from abnormal reverberating circuits. The discharges stimulate many of the neurons to send nerve impulses over their conduction pathways. As a result, lights, noise, or smells may be sensed when the eyes, ears, and nose have not been stimulated. Moreover, the skeletal muscles of a person having a seizure may contract involuntarily. *Partial seizures* begin in a small area on one side of the brain and produce milder symptoms; *generalized seizures* involve larger areas on both sides of the brain and loss of consciousness.

Epilepsy has many causes, including brain damage at birth (the most common cause); metabolic disturbances (hypoglycemia, hypocalcemia, uremia, hypoxia); infections (encephalitis or meningitis); toxins (alcohol, tranquilizers, hallucinogens); vascular disturbances (hemorrhage, hypotension); head injuries; and tumors and abscesses of the brain. Seizures associated with fever are most common in children under the age of two. However, most epileptic seizures have no demonstrable cause.

Epileptic seizures often can be eliminated or alleviated by antiepileptic drugs, such as phenytoin, carbamazepine, and valproate sodium. An implantable device that stimulates the vagus (X) nerve has produced dramatic results in reducing seizures in some patients whose epilepsy was not well-controlled by drugs. In very severe cases, surgical intervention may be an option.

Glaucoma

Glaucoma (glaw-KŌ-ma) is the most common cause of blindness in the United States, afflicting about 2% of the population over age 40. Glaucoma is an abnormally high intraocular pressure due to a buildup of aqueous humor within the anterior cavity. The fluid compresses the lens into the vitreous body and puts pressure on the neurons of the retina. Persistent pressure results in a progression from mild visual impairment to irreversible destruction of neurons of the retina, damage to the optic nerve, and blindness. Glaucoma is painless, and the other eye compensates largely, so a person may experience considerable retinal damage and loss of vision before the condition is diagnosed. Because glaucoma occurs more often with advancing age, regular measurement of intraocular pressure is an increasingly important part of an eye exam as people grow older. Risk factors include race (blacks are more susceptible), increasing age, family history, and past eye injuries and disorders.

Guillain-Barré Syndrome

Guillain-Barré Syndrome *(GBS)* (GĒ-an ba-RĀ) is an acute demyelinating disorder in which macrophages strip myelin from axons in the PNS. It is the most common cause of acute paralysis in North America and Europe and may result from the immune system's response to a bacterial infection. Most patients recover completely or partially, but about 15% remain paralyzed.

Horner's Syndrome

In **Horner's syndrome,** the sympathetic innervation to one side of the face is lost due to an inherited mutation, an injury, or a disease that affects sympathetic outflow through the superior cervical ganglion. Symptoms occur on the affected side and include ptosis (drooping of the upper eyelid), miosis (constricted pupil), and anhidrosis (lack of sweating).

Hydrocephalus

Abnormalities in the brain—tumors, inflammation, or developmental malformations—can interfere with the drainage of CSF from the ventricles into the subarachnoid space. When excess CSF accumulates in the ventricles, the CSF pressure rises. Elevated CSF pressure

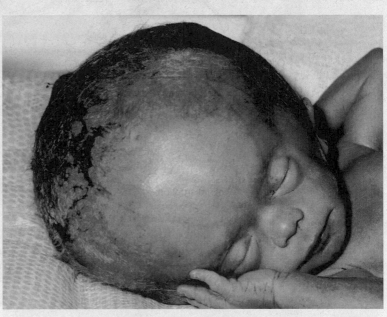

Hydrocephalus in a newborn

causes a condition called **hydrocephalus** (hī'-drō-SEF-a-lus; *hydro-* = water; *cephal-* = head). In a baby whose fontanels have not yet closed, the head bulges due to the increased pressure. If the condition persists, the fluid buildup compresses and damages the delicate nervous tissue. Hydrocephalus is relieved by draining the excess CSF. In one procedure, called *endoscopic third ventriculostomy (ETV)*, a neurosurgeon makes a hole in the floor of the third ventricle and the CSF drains directly into the subarachnoid space. In adults, hydrocephalus may occur after head injury, meningitis, obstruction by cysts or tumors, or subarachnoid hemorrhage. This condition can quickly become life-threatening and requires immediate intervention since the skull bones have already fused.

Hyposmia

Women often have a keener sense of smell than men do, especially at the time of ovulation. Smoking seriously impairs the sense of smell in the short term and may cause long-term damage to olfactory receptors. With aging the sense of smell deteriorates. **Hyposmia** (hī-POZ-mē-a; *osmi* = smell, odor), a reduced ability to smell, affects half of those over age 65 and 75% of those over age 80. Hyposmia also can be caused by neurological changes, such as a head injury, Alzheimer disease, or Parkinson disease; certain drugs, such as antihistamines, analgesics, or steroids; and the damaging effects of smoking.

Ménière's Disease

Ménière's disease (men'-ē-ĀRZ) results from an increased amount of endolymph that enlarges the membranous labyrinth. Among the symptoms are fluctuating hearing loss (caused by distortion of the basilar membrane of the cochlea) and roaring tinnitus (ringing). Spinning or whirling vertigo (dizziness) is characteristic of Ménière's disease. Almost total destruction of hearing may occur over a period of years.

Meningitis

Meningitis (men-in-JĪ-tis; *-itis* = inflammation) is the inflammation of the meninges due to an infection, usually caused by a bacterium or virus. Symptoms include fever, headache, stiff neck, vomiting, confusion, lethargy, and drowsiness. Bacterial meningitis is much more serious and is treated with antibiotics. Viral meningitis has no specific treatment. Bacterial meningitis may be fatal if not treated promptly; viral meningitis usually resolves on its own in 1–2 weeks. A vaccine is available to help protect against some types of bacterial meningitis.

Multiple Sclerosis

Multiple sclerosis (MS) is a disease that causes a progressive destruction of myelin sheaths surrounding neurons in the CNS. It afflicts about 350,000 people

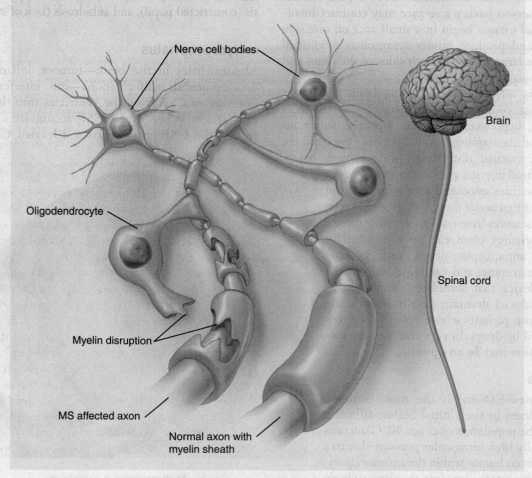

Nerve cell bodies

Brain

Oligodendrocyte

Spinal cord

Myelin disruption

MS affected axon

Normal axon with myelin sheath

Myelin disruption in multiple sclerosis

in the United States and 2 million people worldwide. It usually appears between the ages of 20 and 40, affecting females twice as often as males. MS is most common in whites, less common in blacks, and rare in Asians. MS is an autoimmune disease—the body's own immune system spearheads the attack. The condition's name describes the anatomical pathology: In *multiple* regions the myelin sheaths deteriorate to *scleroses*, which are hardened scars or plaques. Magnetic resonance imaging (MRI) studies reveal numerous plaques in the white matter of the brain and spinal cord. The destruction of myelin sheaths slows and then short-circuits propagation of nerve impulses.

The most common form of the condition is relapsing–remitting MS, which usually appears in early adulthood. The first symptoms may include a feeling of heaviness or weakness in the muscles, abnormal sensations, or double vision. An attack is followed by a period of remission during which the symptoms temporarily disappear. One attack follows another over the years, usually every year or two. The result is a progressive loss of function interspersed with remission periods, during which symptoms abate.

Although the cause of MS is unclear, both genetic susceptibility and exposure to some environmental factor (perhaps a herpes virus) appear to contribute. Since 1993, many patients with relapsing–remitting MS have been treated with injections of beta interferon. This treatment lengthens the time between relapses, decreases the severity of relapses, and slows formation of new lesions in some cases. Unfortunately, not all MS patients can tolerate beta interferon, and therapy becomes less effective as the disease progresses.

■ Otitis Media

Otitis media is an acute infection of the middle ear caused mainly by bacteria and associated with infections of the nose and throat. Symptoms include pain, malaise, fever, and a reddening and outward bulging of the eardrum, which may rupture unless prompt treatment is received. (This may involve draining pus from the middle ear.) Bacteria passing into the auditory tube from the nasopharynx are the primary cause of middle ear infections. Children are more susceptible than adults to middle ear infections because their auditory tubes are almost horizontal, which decreases drainage. If otitis media occurs frequently, a surgical procedure called **tympanotomy** (tim′-pa-NOT-ō-mē; *tympano-* = drum; *-tome* = incision) is often employed. This consists of the insertion of a small tube into the eardrum to provide a pathway for the drainage of fluid from the middle ear.

■ Parkinson Disease

Parkinson disease (PD) is a progressive disorder of the CNS that typically affects its victims around age 60. Neurons that extend from the substantia nigra to the putamen and caudate nucleus, where they release the neurotransmitter dopamine (DA), degenerate in PD. The caudate nucleus of the basal

ganglia contains neurons that liberate the neurotransmitter acetylcholine (ACh). Although the level of ACh does not change as the level of DA declines, the imbalance of neurotransmitter activity—too little DA and too much ACh—is thought to cause most of the symptoms. The cause of PD is unknown, but toxic environmental chemicals, such as pesticides, herbicides, and carbon monoxide, are suspected contributing agents. Only 5% of PD patients have a family history of the disease.

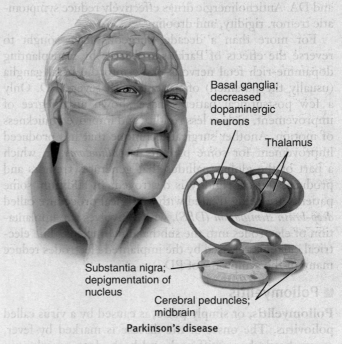

Basal ganglia; decreased dopaminergic neurons

Thalamus

Substantia nigra; depigmentation of nucleus

Cerebral peduncles; midbrain

Parkinson's disease

In PD patients, involuntary skeletal muscle contractions often interfere with voluntary movement. For instance, the muscles of the upper limb may alternately contract and relax, causing the hand to shake. This shaking, called **tremor,** is the most common symptom of PD. Also, muscle tone may increase greatly, causing rigidity of the involved body part. Rigidity of the facial muscles gives the face a masklike appearance. The expression is characterized by a wide-eyed, unblinking stare and a slightly open mouth with uncontrolled drooling.

Motor performance is also impaired by **bradykinesia** (*brady-* = slow), slowness of movements. Activities such as shaving, cutting food, and buttoning a blouse take longer and become increasingly more difficult as the disease progresses. Muscular movements also exhibit **hypokinesia** (*hypo-* = under), decreasing range of motion. For example, words are written smaller, letters are poorly formed, and eventually handwriting becomes illegible. Often, walking is impaired; steps become shorter and shuffling, and arm swing diminishes. Even speech may be affected.

Treatment of PD is directed toward increasing levels of DA and decreasing levels of ACh. Although people with PD do not manufacture enough dopamine, taking it orally is useless because DA cannot cross the blood–brain barrier. Even though symptoms are partially relieved by a drug developed in the 1960s called levodopa (L-dopa), a precursor of DA, the drug does not slow the progression of the disease. As more and more affected brain cells die, the drug becomes useless. Another drug called selegiline (Deprenyl® is used to inhibit monoamine oxidase, an enzyme that degrades catecholamine neurotransmitters such as dopamine. This drug slows progression of PD and may be used together with levodopa. Anticholinergic drugs such as benzotropine and trihexyphenidyl can also be used to block the effects of ACh at some of the synapses between basal ganglia neurons, which helps to restore the balance between ACh and DA. Anticholinergic drugs effectively reduce symptomatic tremor, rigidity, and drooling.

For more than a decade, surgeons have sought to reverse the effects of Parkinson disease by transplanting dopamine-rich fetal nervous tissue into the basal ganglia (usually the putamen) of patients with severe PD. Only a few postsurgical patients have shown any degree of improvement, such as less rigidity and improved quickness of motion. Another surgical technique that has produced improvement for some patients is *pallidotomy*, in which a part of the globus pallidus that generates tremors and produces muscle rigidity is destroyed. In addition, some patients are being treated with a surgical procedure called *deep-brain stimulation (DBS)*, which involves the implantation of electrodes into the subthalamic nucleus. The electrical currents released by the implanted electrodes reduce many of the symptoms of PD.

Poliomyelitis

Poliomyelitis, or simply **polio,** is caused by a virus called poliovirus. The onset of the disease is marked by fever, severe headache, a stiff neck and back, deep muscle pain and weakness, and loss of certain somatic reflexes. In its most serious form, the virus produces paralysis by destroying cell bodies of motor neurons, specifically those in the anterior horns of the spinal cord and in the nuclei of the cranial nerves. Polio can cause death from respiratory or heart failure if the virus invades neurons in vital centers that control breathing and heart functions in the brain stem. Even though polio vaccines have virtually eradicated polio in the United States, outbreaks of polio continue throughout the world. Due to international travel, polio could be easily reintroduced into North America if individuals are not vaccinated appropriately.

Several decades after suffering a severe attack of polio and following their recovery from it, some individuals develop a condition called **post-polio syndrome.** This neurological disorder is characterized by progressive muscle weakness, extreme fatigue, loss of function, and pain, especially in muscles and joints. Post-polio syndrome seems to involve a slow degeneration of motor neurons that innervate muscle fibers. Triggering factors appear to be a fall, a minor accident, surgery, or prolonged bed rest. Possible causes include overuse of surviving motor neurons over time, smaller motor neurons because of the initial infection by the virus, reactivation of dormant polio viral particles, immune-mediated responses, hormone deficiencies, and environmental toxins. Treatment consists of muscle-

strengthening exercises, administration of pyridostigmine to enhance the action of acetylcholine in stimulating muscle contraction, and administration of nerve growth factors to stimulate both nerve and muscle growth.

Presbyopia

With aging, the lens loses elasticity and thus its ability to curve to focus on objects that are close. Therefore, older people cannot read print at the same close range as can younger people. This condition is called **presbyopia** (prez-bē-Ō-pē-a; *presby-* = old; *-opia* = pertaining to the eye or vision). By age 40 the near point of vision may have increased to 20 cm (8 in.), and at age 60 it may be as much as 80 cm (31 in.). Presbyopia usually begins in the mid-forties. At about that age, people who have not previously worn glasses begin to need them for reading. Those who already wear glasses typically start to need bifocals, lenses that can focus for both distant and close vision.

Rabies

Rabies (RĀ-bēz; *rabi-* = mad, raving) is a fatal disease caused by a virus that reaches the CNS via fast axonal transport. It is usually transmitted by the bite of an infected dog or other meat-eating animal. The symptoms are excitement, aggressiveness, and madness, followed by paralysis and death.

Raynaud's Phenomenon

In **Raynaud's phenomenon** (rā-NŌZ) the digits (fingers and toes) become ischemic (lack blood) after exposure to cold or with emotional stress. The condition is due to excessive sympathetic stimulation of smooth muscle in the arterioles of the digits and a heightened response to stimuli that cause vasoconstriction. When arterioles in the digits vasoconstrict in response to sympathetic stimulation, blood flow is greatly diminished. As a result, the digits may blanch (look white due to blockage of blood flow) or become cyanotic (look blue due to deoxygenated blood in capillaries). In extreme cases, the digits may become necrotic from lack of oxygen and nutrients. With rewarming after cold exposure, the arterioles may dilate, causing the fingers and toes to look red. Many patients with Raynaud's phenomenon have low blood pressure. Some have increased numbers of alpha-adrenergic receptors. Raynaud's is most common in young women and occurs more often in cold climates. Patients with Raynaud's phenomenon should avoid exposure to cold, wear warm clothing, and keep the hands and feet warm. Drugs used to treat Raynaud's include nifedipine, a calcium channel blocker that relaxes vascular smooth muscle, and prazosin, which relaxes smooth muscle by blocking alpha receptors. Smoking and the use of alcohol or illicit drugs can exacerbate the symptoms of this condition.

Reflex sympathetic dystrophy

Reflex sympathetic dystrophy (RSD) A syndrome that includes spontaneous pain, painful hypersensitivity to stimuli such as light touch, and excessive coldness and sweating in the involved body part. The disorder frequently involves the forearms, hands, knees, and feet. It appears that activation of the sympathetic division of the autonomic nervous system due to traumatized nociceptors as a result of trauma or surgery on bones or joints is involved. Treatment consists of anesthetics and physical therapy. Recent clinical studies also suggest that the drug baclofen can be used to reduce pain and restore normal function to the affected body part. Also called *complex regional pain syndrome type 1*.

▪ Reye's syndrome

Reye's syndrome (RĪZ) Occurs after a viral infection, particularly chickenpox or influenza, most often in children or teens who have taken aspirin; characterized by vomiting and brain dysfunction (disorientation, lethargy, and personality changes) that may progress to coma and death.

▪ Shingles

Shingles is an acute infection of the peripheral nervous system caused by herpes zoster (HER-pēz ZOS-ter), the virus that also causes chickenpox. After a person recovers from chickenpox, the virus retreats to a posterior root ganglion. If the virus is reactivated, the immune system usually prevents it from spreading. From time to time, however, the reactivated virus overcomes a weakened immune system, leaves the ganglion, and travels down sensory neurons of the skin by fast axonal transport (described on page 419). The result is pain, discoloration of the skin, and a characteristic line of skin blisters. The line of blisters marks the distribution (dermatome) of the particular cutaneous sensory nerve belonging to the infected posterior root ganglion.

▪ Syphilis

Syphilis is a sexually transmitted disease caused by the bacterium *Treponema pallidum*. Because it is a bacterial infection, it can be treated with antibiotics. However, if the infection is not treated, the third stage of syphilis typically causes debilitating neurological symptoms. A common outcome is progressive degeneration of the posterior portions of the spinal cord, including the posterior columns, posterior spinocerebellar tracts, and posterior roots. Somatic sensations are lost, and the person's gait becomes uncoordinated and jerky because proprioceptive impulses fail to reach the cerebellum.

▪ Transient Ischemic Attacks

A **transient ischemic attack (TIA)** is an episode of temporary cerebral dysfunction caused by impaired blood flow to the brain. Symptoms include dizziness, weakness, numbness, or paralysis in a limb or in one side of the body; drooping of one side of the face; headache; slurred speech or difficulty understanding speech; and a partial loss of vision or double vision. Sometimes nausea or vomiting also occurs. The onset of symptoms is sudden and reaches maximum intensity almost immediately. A TIA usually persists for 5 to 10 minutes and only rarely lasts as long as 24 hours. It leaves no permanent neurological deficits. The causes of the impaired blood flow that lead to TIAs are blood clots, atherosclerosis, and certain blood disorders. About one-third of patients who experience a TIA will have a CVA eventually. Therapy for TIAs includes drugs such as aspirin, which blocks the aggregation of blood platelets, and anticoagulants; cerebral artery bypass grafting; and carotid endarterectomy (removal of the cholesterol-containing plaques and inner lining of an artery).

Additional Clinical Considerations

▪ Aphasia

Much of what we know about language areas comes from studies of patients with language or speech disturbances that have resulted from brain damage. Broca's speech area, Wernicke's area, and other language areas are located in the left cerebral hemisphere of most people, regardless of whether they are left-handed or right-handed. Injury to language areas of the cerebral cortex results in **aphasia** (a-FĀ-zē-a; *a-* = without; *-phasia* = speech), an inability to use or comprehend words. Damage to Broca's speech area results in *nonfluent aphasia*, an inability to properly articulate or form words; people with nonfluent aphasia know what they wish to say but cannot speak. Damage to Wernicke's area, the common integrative area, or auditory association area results in *fluent aphasia*, characterized by faulty understanding of spoken or written words. A person experiencing this type of aphasia may fluently produce strings of words that have no meaning ("word salad"). For example, someone with fluent aphasia might say, "I rang car porch dinner light river pencil." The underlying deficit may be **word deafness** (an inability to understand spoken words), **word blindness** (an inability to understand written words), or both.

▪ Ataxia

Damage to the cerebellum through trauma or disease disrupts muscle coordination, a condition called **ataxia** (*a-* = without; *-taxia* = order). Blindfolded people with ataxia cannot touch the tip of their nose with a finger because they cannot coordinate movement with their sense of where a body part is located. Another sign of ataxia is changed speech pattern due to uncoordinated speech muscles. Cerebellar damage may also result in staggering or abnormal walking movements. People who consume too much alcohol show signs of ataxia because alcohol inhibits activity of the cerebellum.

▪ Autonomic Dysreflexia

Autonomic dysreflexia (dis′-rē-FLEX-sē-a) is an exaggerated response of the sympathetic division of the ANS that occurs in about 85% of individuals with spinal cord injury at or above the level of T6. The condition is seen after recovery from spinal shock (see page 489) and occurs due to interruption of the control of ANS neurons by higher centers. When certain sensory impulses, such as those resulting from stretching of a full urinary bladder, are unable to ascend the spinal cord, mass stimulation of the sympathetic nerves inferior to the level of injury occurs. Other triggers include stimulation of pain receptors and the visceral contractions resulting from sexual stimulation, labor/delivery, and bowel stimulation. Among the effects of increased sympathetic activity is severe vasoconstriction, which elevates blood pressure. In response, the cardiovascular center in the medulla oblongata (1) increases parasympathetic output via the vagus (X) nerve, which decreases heart rate, and (2) decreases sympathetic output, which causes dilation of blood vessels superior to the level of the injury.

Autonomic dysreflexia is characterized by a pounding headache; hypertension; flushed, warm skin with profuse sweating above the injury level; pale, cold, and dry skin below the injury level; and anxiety. It is an emergency condition that requires immediate intervention. The first approach is to quickly identify the problematic stimulus and remove it. If this does not relieve the symptoms, an antihypertensive drug such as clonidine or nitroglycerine can be administered. If left untreated, autonomic dysreflexia can cause seizures, stroke, or heart attack.

■ Injuries to Nerves Emerging from the Brachial Plexus

Injury to the superior roots of the brachial plexus (C5–C6) may result from forceful pulling away of the head from the shoulder, as might occur from a heavy fall on the shoulder or excessive stretching of an infant's neck during childbirth. The presentation of this injury is characterized by an upper limb in which the shoulder is adducted, the arm is medially rotated, the elbow is extended, the forearm is pronated, and the wrist is flexed. This condition is called **Erb-Duchenne palsy** or **waiter's tip position.** There is loss of sensation along the lateral side of the arm.

Radial (and axillary) **nerve injury** can be caused by improperly administered intramuscular injections into the deltoid muscle. The radial nerve may also be injured when a cast is applied too tightly around the mid-humerus. Radial nerve injury is indicated by **wrist drop,** the inability to extend the wrist and fingers. Sensory loss is minimal due to the overlap of sensory innervation by adjacent nerves.

Median nerve injury may result in **median nerve palsy,** which is indicated by numbness, tingling, and pain in the palm and fingers. There is also inability to pronate the forearm and flex the proximal interphalangeal joints of all digits and the distal interphalangeal joints of the second and third digits. In addition, wrist flexion is weak and is accompanied by adduction, and thumb movements are weak.

Ulnar nerve injury may result in **ulnar nerve palsy,** which is indicated by an inability to abduct or adduct the fingers, atrophy of the interosseous muscles of the hand, hyperextension of the metacarpophalangeal joints, and flexion of the interphalangeal joints, a condition called **clawhand.** There is also loss of sensation over the little finger.

Long thoracic nerve injury results in paralysis of the serratus anterior muscle. The medial border of the scapula protrudes, giving it the appearance of a wing. When the arm is raised, the vertebral border and inferior angle of the scapula pull away from the thoracic wall and protrude outward, causing the medial border of the scapula to protrude; because the scapula looks like a wing, this condition is called **winged scapula.** The arm cannot be abducted beyond the horizontal position.

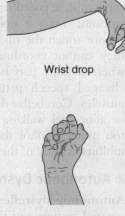

Wrist drop

Erb-Duchenne palsy
(waiter's tip)

Median nerve palsy

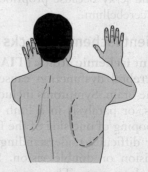

Ulnar nerve palsy   Winging of right scapula

■ Brain Injuries

Brain injuries are commonly associated with head trauma and result in part from displacement and distortion of neural tissue at the moment of impact. Additional tissue damage may occur when normal blood flow is restored after a period of ischemia (reduced blood flow). The sudden increase in oxygen level produces large numbers of oxygen free radicals (charged oxygen molecules with an unpaired electron). Brain cells recovering from the effects of a stroke or cardiac arrest also release free radicals. Free radicals cause damage by disrupting cellular DNA and enzymes and by altering plasma membrane permeability. Brain injuries can also result from hypoxia (oxygen deprivation).

Various degrees of brain injury are described by specific terms. A **concussion** is an injury characterized by an abrupt, but temporary, loss of consciousness (from seconds to hours), disturbances of vision, and problems with equilibrium. It is caused by a blow to the head or the sudden stopping of a moving head (as in an automobile accident) and is the most common brain injury. A concussion produces no obvious bruising of the brain. Signs of a concussion are headache, drowsiness, nausea and/or vomiting, lack of concentration, confusion, or post-traumatic amnesia (memory loss).

A brain **contusion** is bruising due to trauma and includes the leakage of blood from microscopic vessels. It is usually associated with a concussion. In a contusion, the pia mater may be torn, allowing blood to enter the subarachnoid space. The area most commonly affected is the frontal lobe. A contusion usually results in an immediate loss of consciousness (generally lasting no longer than 5 minutes), loss of reflexes, transient cessation of respiration, and decreased blood pressure. Vital signs typically stabilize in a few seconds.

A **laceration** is a tear of the brain, usually from a skull fracture or a gunshot wound. A laceration results in rupture of large blood vessels, with bleeding into the brain and subarachnoid space. Consequences include cerebral hematoma (localized pool of blood, usually clotted, that swells against the brain tissue), edema, and increased intracranial pressure. If the blood clot is small enough, it may pose no major threat and may be absorbed. If the blood clot is large, it may require surgical evacuation. Swelling infringes on the limited space that the brain occupies in the cranial cavity. Swelling causes excruciating headaches. Brain tissue can become necrotic (die) due to the swelling; if the swelling is severe enough, the brain can herniate through the foramen magnum, resulting in death.

■ Brain Tumors

A **brain tumor** is an abnormal growth of tissue in the brain that may be malignant or benign. Unlike most other tumors in the body, malignant and benign tumors may be equally serious, compressing adjacent tissues and causing a buildup of pressure in the skull. The most common malignant tumors are secondary tumors that metastasize from other cancers in the body, such as those in the lungs, breasts, skin (malignant melanoma), blood (leukemia), and lymphatic organs (lymphoma). Most primary brain tumors (those that originate within the brain) are gliomas, which develop in neuroglia. The symptoms of a brain tumor depend on its size, location, and rate of growth. Among the symptoms are headache, poor balance and coordination, dizziness, double vision, slurred speech, nausea and vomiting, fever, abnormal pulse and breathing rates, personality changes, numbness and weakness of the limbs, and seizures. Treatment options for brain tumors vary with their size, location, and type and may include surgery, radiation therapy, and/or chemotherapy. Unfortunately, chemotherapeutic agents do not readily cross the blood–brain barrier.

■ Cataracts

A common cause of blindness is a loss of transparency of the lens known as a **cataract** (CAT-a-rakt = waterfall). The lens becomes cloudy (less transparent) due to changes in the structure of the lens proteins. Cataracts often occur with aging but may also be caused by injury, excessive exposure to ultraviolet rays, certain medications (such as long-term use of steroids), or complications of other diseases (for example, diabetes). People who smoke also have increased risk of developing cataracts. Fortunately, sight can usually be restored by surgical removal of the old lens and implantation of a new artificial one.

■ Color Blindness and Night Blindness

Most forms of **color blindness**, an inherited inability to distinguish between certain colors, result from the absence or deficiency of one of the three types of cones. The most common type is *red-green color blindness*, in which red cones or green cones are missing. As a result, the person cannot distinguish between red and green. Prolonged vitamin A deficiency and the resulting below-normal amount of rhodopsin may cause **night blindness** or **nyctalopia** (nik'-ta-LŌ-pē-a), an inability to see well at low light levels.

■ Deafness

Deafness is significant or total hearing loss. **Sensorineural deafness** is caused by either impairment of hair cells in the cochlea or damage of the cochlear branch of the vestibulocochlear (VIII) nerve. This type of deafness may be caused by atherosclerosis, which reduces blood supply to the ears; by repeated exposure to loud noise, which destroys hair cells of the spiral organ; and/or by certain drugs such as aspirin and streptomycin. **Conduction deafness** is caused by impairment of the external and middle ear mechanisms for transmitting sounds to the cochlea. Causes of conduction deafness include otosclerosis, the deposition of new bone around the oval window; impacted cerumen; injury to the eardrum; and aging, which often results in thickening of the eardrum and stiffening of the joints of the auditory ossicles. A hearing test called *Weber's test* is used to distinguish between sensorineural and conduction deafness. In the test, the stem of a vibrating fork is held to the forehead. In people with normal hearing, the sound is heard equally in both ears. If the sound is heard best in the affected ear, the deafness is probably of the conduction type; if the sound is heard best in the normal ear, it is probably of the sensorineural type.

Detached Retina

A **detached retina** may occur due to trauma, such as a blow to the head, in various eye disorders, or as a result of age-related degeneration. The detachment occurs between the neural portion of the retina and the pigment epithelium. Fluid accumulates between these layers, forcing the thin, pliable retina to billow outward. The result is distorted vision and blindness in the corresponding field of vision. The retina may be reattached by laser surgery or cryosurgery (localized application of extreme cold) and reattachment must be accomplished quickly to avoid permanent damage to the retina.

Excitotoxicity

A high level of glutamate in the interstitial fluid of the CNS causes **excitotoxicity**—destruction of neurons through prolonged activation of excitatory synaptic transmission. The most common cause of excitotoxicity is oxygen deprivation of the brain due to ischemia (inadequate blood flow), as happens during a stroke. Lack of oxygen causes the glutamate transporters to fail, and glutamate accumulates in the interstitial spaces between neurons and glia, literally stimulating the neurons to death. Clinical trials are underway to see if antiglutamate drugs administered after a stroke can offer some protection from excitotoxicity.

Loud Sounds and Hair Cell Damage

Exposure to loud music and the engine roar of jet planes, revved-up motorcycles, lawn mowers, and vacuum cleaners damages hair cells of the cochlea. Because prolonged noise exposure causes hearing loss, employers in the United States must require workers to use hearing protectors when occupational noise levels exceed 90 dB. Rock concerts and even inexpensive headphones can easily produce sounds over 110 dB. Continued exposure to high-intensity sounds is one cause of **deafness,** a significant or total hearing loss. The louder the sounds, the more rapid is the hearing loss. Deafness usually begins with loss of sensitivity for high-pitched sounds. If you are listening to music through head-phones and bystanders can hear it, the dB level is in the damaging range. Most people fail to notice their progressive hearing loss until destruction is extensive and they begin having difficulty understanding speech. Wearing earplugs with a noise-reduction rating of 30 dB while engaging in noisy activities can protect the sensitivity of your ears.

Lumbar Plexus Injuries

The largest nerve arising from the lumbar plexus is the femoral nerve. **Femoral nerve injury,** which can occur in stab or gunshot wounds, is indicated by an inability to extend the leg and by loss of sensation in the skin over the anteromedial aspect of the thigh.

Obturator nerve injury results in paralysis of the adductor muscles of the thigh and loss of sensation over the medial aspect of the thigh. It may result from pressure on the nerve by the fetal head during pregnancy.

Injury of the Medulla

Given the vital activities controlled by the medulla, it is not surprising that **injury to the medulla** from a hard blow to the back of the head or upper neck such as falling back on ice can be fatal. Damage to the medullary rhythmicity area is particularly serious and can rapidly lead to death. Symptoms of nonfatal injury to the medulla may include cranial nerve malfunctions on the same side of the body as the injury, paralysis and loss of sensation on the opposite side of the body, and irregularities in breathing or heart rhythm. Alcohol overdose also suppresses the medullary rhythmicity area and may result in death.

Paralysis

Damage or disease of *lower* motor neurons produces **flaccid paralysis** of muscles on the same side of the body. There is neither voluntary nor reflex action of the innervated muscle fibers, muscle tone is decreased or lost, and the muscle remains limp or flaccid. Injury or disease of *upper* motor neurons in the cerebral cortex removes inhibitory influences that some of these neurons have on lower motor neurons, which causes **spastic paralysis** of muscles on the opposite side of the body. In this condition muscle tone is increased, reflexes are exaggerated, and pathological reflexes such as the Babinski sign appear.

Phantom Limb Sensation

Patients who have had a limb amputated may still experience sensations such as itching, pressure, tingling, or pain as if the limb were still there. This phenomenon is called **phantom limb sensation.** One explanation for phantom limb sensations is that the cerebral cortex interprets impulses arising in the proximal portions of sensory neurons that previously carried impulses from the limb as coming from the nonexistent (phantom) limb. Another explanation for phantom limb sensations is that the brain itself contains networks of neurons that generate sensations of body awareness. In the latter view, neurons in the brain that previously received sensory impulses from the missing limb are still active, giving rise to false sensory perceptions. Phantom limb pain can be very distressing to an amputee. Many report that the pain is severe or extremely intense, and that it often does not respond to traditional pain medication therapy. In such cases, alternative treatments may include electrical nerve stimulation, acupuncture, and biofeedback.

Injuries to the Phrenic Nerves

Complete severing of the spinal cord above the origin of the phrenic nerves (C3, C4, and C5) causes respiratory arrest. Breathing stops because the phrenic nerves no longer send nerve impulses to the diaphragm. The phrenic nerves may also be damaged due to pressure from malignant tumors in the mediastinum.

■ Sciatic Nerve Injury

The most common form of back pain is caused by compression or irritation of the sciatic nerve, the longest nerve in the human body. Injury to the sciatic nerve and its branches results in **sciatica,** pain that may extend from the buttock down the posterior and lateral aspect of the leg and the lateral aspect of the foot. The sciatic nerve may be injured because of a herniated (slipped) disc, dislocated hip, osteoarthritis of the lumbosacral spine, pressure from the uterus during pregnancy, inflammation, irritation, or an improperly administered gluteal intramuscular injection.

In the majority of sciatic nerve injuries, the common fibular portion is the most affected, frequently from fractures of the fibula or by pressure from casts or splints. Damage to the common fibular nerve causes the foot to be plantar flexed, a condition called **footdrop,** and inverted, a condition called **equinovarus.** There is also loss of function along the anterolateral aspects of the leg and dorsum of the foot and toes. Injury to the tibial portion of the sciatic nerve results in dorsiflexion of the foot plus eversion, a condition called **calcaneovalgus.** Loss of sensation on the sole also occurs. Treatments for sciatica are similar to those outlined earlier for a herniated (slipped) disc—rest, pain medications, exercises, ice or heat, and massage.

■ Spinal Cord Compression

Although the spinal cord is normally protected by the vertebral column, certain disorders may put pressure on it and disrupt its normal functions. Spinal cord compression may result from fractured vertebrae, herniated intervertebral discs, tumors, osteoporosis, or infections. If the source of the compression is determined before neural tissue is destroyed, spinal cord function usually returns to normal. Depending on the location and degree of compression, symptoms include pain, weakness or paralysis, and either decreased or complete loss of sensation below the level of the injury.

■ Spinal Cord Injury

Most spinal cord injuries are due to trauma as a result of factors such as automobile accidents, falls, contact sports, diving, or acts of violence. The effects of the injury depend on the extent of direct trauma to the spinal cord or compression of the cord by fractured or displaced vertebrae or blood clots. Although any segment of the spinal cord may be involved, most common sites of injury are in the cervical, lower thoracic, and upper lumbar regions. Depending on the location and extent of spinal cord damage, paralysis may occur. **Monoplegia** (*mono-* = one; *-plegia* = blow or strike) is paralysis of one limb only. **Diplegia** (*di-* = two) is paralysis of both upper limbs or both lower limbs. **Paraplegia** (*para-* = beyond) is paralysis of both lower limbs. **Hemiplegia** (*hemi-* = half) is paralysis of the upper limb, trunk, and lower limb on one side of the body, and **quadriplegia** (*quad-* = four) is paralysis of all four limbs.

Complete transection (tran-SEK-shun; *trans-* = across; *-section* = a cut) of the spinal cord means that the cord is severed from one side to the other, thus cutting all sensory and motor tracts. It results in a loss of all sensations and voluntary movement *below* the level of the transection. A person will have permanent loss of all sensations in dermatomes below the injury because ascending nerve impulses cannot propagate past the transection to reach the brain. At the same time, voluntary muscle contractions will be lost below the transection because nerve impulses descending from the brain also cannot pass. The extent of paralysis of skeletal muscles depends on the level of injury. The following list outlines which muscle functions may be *retained* at progressively lower levels of spinal cord transection.

- C1–C3: no function maintained from the neck down; ventilator needed for breathing
- C4–C5: diaphragm, which allows breathing
- C6–C7: some arm and chest muscles, which allows feeding, some dressing, and propelling wheelchair
- T1–T3: intact arm function
- T4–T9: control of trunk above the umbilicus
- T10–L1: most thigh muscles, which allows walking with long leg braces
- L1–L2: most leg muscles, which allows walking with short leg braces

Hemisection is a partial transection of the cord on either the right or left side. After hemisection, three main symptoms, known together as *Brown-Sequard syndrome* (se-KAR) occur below the level of injury: (1) Damage of the posterior column causes loss of proprioception and touch sensations on the ipsilateral (same) side as the injury. (2) Damage of the lateral corticospinal tract causes ipsilateral paralysis. (3) Damage of the spinothalamic tract causes loss of pain and temperature sensations on the contralateral (opposite) side.

Following complete transection, and to varying degrees after hemisection, spinal shock occurs. **Spinal shock** is an immediate response to spinal cord injury characterized by temporary **areflexia** (a'-rē-FLEK-sē-a), loss of reflex function. The areflexia occurs in parts of the body served by spinal nerves below the level of the injury. Signs of acute spinal shock include slow heart rate, low blood pressure, flaccid paralysis of skeletal muscles, loss of somatic sensations, and urinary bladder dysfunction. Spinal shock may begin within 1 hour after injury and may last from several minutes to several months, after which reflex activity gradually returns.

In many cases of traumatic injury of the spinal cord, the patient may have an improved outcome if an anti-inflammatory corticosteroid drug called methylprednisolone is given within 8 hours of the injury. This is because the degree of neurologic deficit is greatest immediately following traumatic injury as a result of edema (collection of fluid within tissues) as the immune system responds to injury.

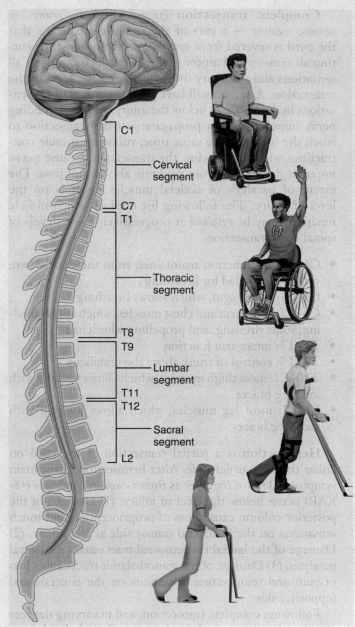

C1

Cervical
segment

C7
T1

Thoracic
segment

T8
T9

Lumbar
segment

T11
T12

Sacral
segment

L2

Spinal cord injuries

■ Strychnine Poisoning

The importance of inhibitory neurons can be appreciated by observing what happens when their activity is blocked. Normally, inhibitory neurons in the spinal cord called *Renshaw cells* release the neurotransmitter glycine at inhibitory synapses with somatic motor neurons. This inhibitory input to their motor neurons prevents excessive contraction of skeletal muscles. Strychnine is a lethal poison that binds to and blocks glycine receptors. The normal, delicate balance between excitation and inhibition in the CNS is disturbed, and motor neurons generate nerve impulses without restraint. All skeletal muscles, including the diaphragm, contract fully and remain contracted. Because the diaphragm cannot relax, the victim cannot inhale, and suffocation results.

■ Taste Aversion

Probably because of taste projections to the hypothalamus and limbic system, there is a strong link between taste and pleasant or unpleasant emotions. Sweet foods evoke reactions of pleasure while bitter ones cause expressions of disgust even in newborn babies. This phenomenon is the basis for **taste aversion,** in which people and animals quickly learn to avoid a food if it upsets the digestive system. The advantage of avoiding foods that cause such illness is longer survival. However, the drugs and radiation treatments used to combat cancer often cause nausea and gastrointestinal upset regardless of what foods are consumed. Thus, cancer patients may lose their appetite because they develop taste aversions for most foods. ■

Medical Terminology

Ageusia (a-GŪ-sē-ā; *a-* = without; *-geusis* = taste) Loss of the sense of taste.

Agnosia (ag-NŌ-zē-a; *a-* = without; *-gnosia* = knowledge) Inability to recognize the significance of sensory stimuli such as sounds, sights, smells, tastes, and touch.

Amblyopia (am′-blē-Ō-pē-a; *ambly-* = dull or dim) Term used to describe the loss of vision in an otherwise normal eye that, because of muscle imbalance, cannot focus in synchrony with the other eye. Sometimes called "wandering eyeball" or a "lazy eye."

Anosmia (an-OZ-mē-a; *a-* = without; *osmi* = smell, odor) Total lack of the sense of smell.

Apraxia (a-PRAK-sē-a; *-praxia* = coordinated) Inability to carry out purposeful movements in the absence of paralysis.

Autonomic nerve neuropathy (noo-ROP-a-thē) If a neuropathy (specifically a disorder of a cranial or spinal nerve) affects one or more autonomic nerves, there can be multiple effects on the autonomic nervous system that interfere with reflexes. These include fainting and low blood pressure when standing (orthostatic hypotension) due to decreased sympathetic control of the cardiovascular system, constipation, urinary incontinence, and impotence. This type of neuropathy is often caused by long-term diabetes mellitus and is known as ***diabetic retinopathy.***

Barotrauma (bar′-ō-TRAW-ma; *baros-* = weight) Damage or pain, mainly affecting the middle ear, as a result of pressure changes. It occurs when pressure on the outer side of the tympanic membrane is higher than on the inner side, for example, when flying in an airplane or diving. Swallowing or holding your nose and exhaling with your mouth closed usually opens the auditory tubes, allowing air into the middle ear to equalize the pressure.

Blepharitis (blef-a-RĪ-tis; *blephar-* = eyelid; *-itis* = inflammation of) An inflammation of the eyelid.

Coma (KŌ-ma) A state of unconsciousness in which a person's responses to stimuli are reduced or absent. In a *light coma*, an individual may respond to certain stimuli, such as sound, touch, or light, and move his or her eyes, cough, and even murmur. In a *deep coma*, a person does not respond to any stimuli and does not make any movements. Causes of coma include head injuries, cardiac arrest, stroke, brain tumors, infections (encephalitis and meningitis), seizures, alcoholic intoxication, drug overdose, severe lung disorders (chronic obstructive pulmonary disease, pulmonary edema, pulmonary embolism), inhalation of large amounts of carbon monoxide, liver or kidney failure, low or high blood sugar or sodium levels, and low or high body temperature. If brain damage is minor or reversible, a person may come out of a coma and recover fully; if brain damage is severe and irreversible, recovery is unlikely.

Conjunctivitis (pinkeye) An inflammation of the conjunctiva; when caused by bacteria such as pneumococci, staphylococci, or *Hemophilus influenzae*, it is very contagious and more common in children. Conjunctivitis may also be caused by irritants, such as dust, smoke, or pollutants in the air, in which case it is not contagious.

Consciousness (KON-shus-nes) A state of wakefulness in which an individual is fully alert, aware, and oriented, partly as a result of feedback between the cerebral cortex and reticular activating system.

Corneal abrasion (KOR-nē-al a-BRĀ-zhun) A scratch on the surface of the cornea, for example, from a speck of dirt or damaged contact lenses. Symptoms include pain, redness, watering, blurry vision, sensitivity to bright light, and frequent blinking.

Delirium (dē-LIR-ē-um = off the track) A transient disorder of abnormal cognition and disordered attention accompanied by disturbances of the sleep–wake cycle and psychomotor behavior (hyperactivity or hypoactivity of movements and speech). Also called **acute confusional state (ACS).**

Dementia (de-MEN-shē-a; *de-* = away from; *-mentia* = mind) Permanent or progressive general loss of intellectual abilities, including impairment of memory, judgment, abstract thinking, and changes in personality.

Demyelination (dē-mī-e-li-NĀ-shun) refers to the loss or destruction of myelin sheaths around axons. It may result from disorders such as multiple sclerosis or Tay-Sachs disease or from medical treatments such as radiation therapy and chemotherapy. Any single episode of demyelination may cause deterioration of affected nerves.

Diabetic retinopathy (ret-i-NOP-a-thē; *retino-* = retina; *-pathos* = suffering) Degenerative disease of the retina due to diabetes mellitus, in which blood vessels in the retina are damaged or new ones grow and interfere with vision.

Dysautonomia (dis-aw-tō-NŌ-mē-a; *dys-* = difficult; *autonomia* = self-governing) An inherited disorder in which the autonomic nervous system functions abnormally, resulting in reduced tear gland secretions, poor vasomotor control, motor incoordination, skin blotching,

absence of pain sensation, difficulty in swallowing, hyporeflexia, excessive vomiting, and emotional instability.

Encephalopathy (en-sef′-a-LOP-a-thē; *encephalo* = brain; *-pathos* 5 disease) Any disorder of the brain.

Epidural block Injection of an anesthetic drug into the epidural space, the space between the dura mater and the vertebral column, in order to cause a temporary loss of sensation. Such injections in the lower lumbar region are used to control pain during childbirth.

Exotropia (ek′-sō-TRŌ-pē-a; *ex-* = out; *-tropia* = turning) Turning outward of the eyes.

Hyperhydrosis (hi′-per-hī-DRŌ-sis; *hyper-* = above or too much; *hidrosis* = sweat; *-osis* = condition) Excessive or profuse sweating due to intense stimulation of sweat glands.

Insomnia (in-SOM-nē-a; *in-* = not; *-somnia* = sleep) Difficulty in falling asleep and staying asleep.

Keratitis (ker′-a-TĪ-tis; *kerat-* = cornea) An inflammation or infection of the cornea.

Lethargy (LETH-ar-jē) A condition of functional sluggishness.

Mass reflex In cases of severe spinal cord injury above the level of the sixth thoracic vertebra, stimulation of the skin or overfilling of a visceral organ (such as the urinary bladder or colon) below the level of the injury results in intense activation of autonomic and somatic output from the spinal cord as reflex activity returns. The exaggerated response occurs because there is no inhibitory input from the brain. The mass reflex consists of flexor spasms of the lower limbs, evacuation of the urinary bladder and colon, and profuse sweating below the level of the lesion.

Megacolon (*mega-* = big) An abnormally large colon. In congenital megacolon, parasympathetic nerves to the distal segment of the colon do not develop properly. Loss of motor function in the segment causes massive dilation of the normal proximal colon. The condition results in extreme constipation, abdominal distension, and occasionally, vomiting. Surgical removal of the affected segment of the colon corrects the disorder.

Microcephaly (mī-krō-Sef-a-lē; *micro-* = small; *-cephal* = head) A congenital condition that involves the development of a small brain and skull and frequently results in mental retardation.

Miosis (mī-Ō-sis) Constriction of the pupil.

Motion sickness Paleness, restlessness, nausea, weakness, dizziness, and malaise that may progress to vomiting caused by increased activity of the semicircular canals. It occurs during motion, for example, in a car, on a boat, on a train, or in an airplane. Usually, when motion stops, symptoms improve. Over-the-counter medications such as meclizine (Bonine®) or dimenhydrinate (Dramamine®) can be taken before embarking on a trip. A prescription skin patch that contains scopolamine (Transderm Scop®) can also be taken before symptoms occur.

Mydriasis (mi-DRĪ-a-sis) Dilation of the pupil.

Myelitis (mī-e-LĪ-tis; *myel-* = spinal cord) Inflammation of the spinal cord.

Narcolepsy (NAR-kō-lep-sē; *narco-* = numbness; *-lepsy* = a seizure) A condition in which REM sleep cannot be inhibited during waking periods. As a result, involuntary periods of sleep that last about 15 minutes occur throughout the day.

Nerve block Loss of sensation in a region due to injection of a local anesthetic; an example is local dental anesthesia.

Neuralgia (noo-RAL-jē-a; *neur-* = nerve; *-algia* = pain) Attacks of pain along the entire course or a branch of a sensory nerve.

Neuritis (*neur-* = nerve; *-itis* = inflammation) Inflammation of one or several nerves that may result from irritation to the nerve produced by direct blows, bone fractures, contusions, or penetrating injuries. Additional causes include infections, vitamin deficiency (usually thiamine), and poisons such as carbon monoxide, carbon tetrachloride, heavy metals, and some drugs.

Neuroblastoma (noor-ō-blas-TŌ-ma) A malignant tumor that consists of immature nerve cells (neuroblasts); occurs most commonly in the abdomen and most frequently in the adrenal glands. Although rare, it is the most common tumor in infants.

Neuropathy (noo-ROP-a-thē; *neuro-* = a nerve; *-pathy* = disease) Any disorder that affects the nervous system but particularly a disorder of a cranial or spinal nerve. An example is *facial neuropathy* (Bell's palsy), a disorder of the facial (VII) nerve.

Nystagmus (nis-TAG-mus; *nystagm-* = nodding or drowsy) A rapid involuntary movement of the eyeballs, possibly caused by a disease of the central nervous system. It is associated with conditions that cause vertigo.

Otalgia (ō-TAL-jē-a; *oto-* = ear; *-algia* = pain) Earache.

Pain threshold The smallest intensity of a painful stimulus at which a person perceives pain. All individuals have the same pain threshold.

Pain tolerance The greatest intensity of painful stimulation that a person is able to tolerate. Individuals vary in their tolerance to pain.

Paresthesia (par-es-THĒ-zē-a; *par-* = departure from normal; *-esthesia* = sensation) An abnormal sensation such as burning, pricking, tickling, or tingling resulting from a disorder of a sensory nerve.

Photophobia (fō′-tō-FŌ-bē-a; *photo-* = light; *-phobia* = fear) Abnormal visual intolerance to light.

Prosopagnosia (pros′-ō-pag-NŌ-sē-a; *a-* = without; *-gnosia* = knowledge) Inability to recognize faces, usually caused by damage to the facial recognition area in the inferior temporal lobe of both cerebral hemispheres.

Ptosis (TŌ-sis = fall) Falling or drooping of the eyelid (or slippage of any organ below its normal position).

Retinoblastoma (ret-i-nō-blas-TŌ-ma; *-oma* = tumor) A tumor arising from immature retinal cells; it accounts for 2% of childhood cancers.

Scotoma (skō-TŌ-ma = darkness) An area of reduced or lost vision in the visual field.

Sleep apnea (AP-nē -a; *a-* = without; *-pnea* = breath) A disorder in which a person repeatedly stops breathing for 10 or more seconds while sleeping. Most often, it occurs because loss of muscle tone in pharyngeal muscles allows the airway to collapse.

Strabismus (stra-BIZ-mus; *strabismos* = squinting) Misalignment of the eyeballs so that the eyes do not move in unison when viewing an object; the affected eye turns either medially or laterally with respect to the normal eye and the result is double vision (diplopia). It may be caused by physical trauma, vascular injuries, or tumors of the extrinsic eye muscle or the oculomotor (III), trochlear (IV), or abducens (VI) cranial nerves.

Stupor (STOO-por) Unresponsiveness from which a patient can be aroused only briefly and only by vigorous and repeated stimulation.

Synesthesia (sin-es-THĒ-zē-a; *syn-* = together; *aisthesis* = sensation) A condition in which sensations of two or more modalities accompany one another. In some cases, a stimulus for one sensation is perceived as a stimulus for another; for example, a sound produces a sensation of color. In other cases, a stimulus from one part of the body is experienced as coming from a different part.

Tinnitus (ti-NĪ-tus) A ringing, roaring, or clicking in the ears.

Tonometer (tō-NOM-ē-ter; *tono-* = tension or pressure; *-metron* = measure) An instrument for measuring pressure, especially intraocular pressure.

Trachoma (tra-KŌ-ma) A serious form of conjunctivitis and the greatest single cause of blindness in the world. It is caused by the bacterium *Chlamydia trachomatis*. The disease produces an excessive growth of subconjunctival tissue and invasion of blood vessels into the cornea, which progresses until the entire cornea is opaque.

Vagotomy (vā-GOT-ō-mē; *-tome* = incision) Cutting the vagus (X) nerve. It is frequently done to decrease the production of hydrochloric acid in persons with ulcers.

Vertigo (VER-ti-gō = dizziness) A sensation of spinning or movement in which the world seems to revolve or the person seems to revolve in space, often associated with nausea and, in some cases, vomiting. It may be caused by arthritis of the neck or an infection of the vestibular apparatus.

Chapter 8 | The Endocrine System

The endocrine system controls body activities by releasing mediators, called **hormones**. A hormone is released in one part of the body but regulates the activity of cells in other parts of the body. *Circulating hormones* pass from the secretory cells that make them into interstitial fluid and then into the blood. *Local hormones* act locally on neighboring cells or on the same cell that secreted them without first entering the bloodstream. Several hormones regulate metabolism, uptake of glucose, and molecules used for ATP production by body cells. Disorders of the endocrine system often involve either *hyposecretion*, inadequate release of a hormone, or *hypersecretion*, excessive release of a hormone. Sometimes the problem is faulty hormone receptors, an inadequate number of receptors, or defects in second-messenger systems. Problems associated with endocrine dysfunction can be widespread.

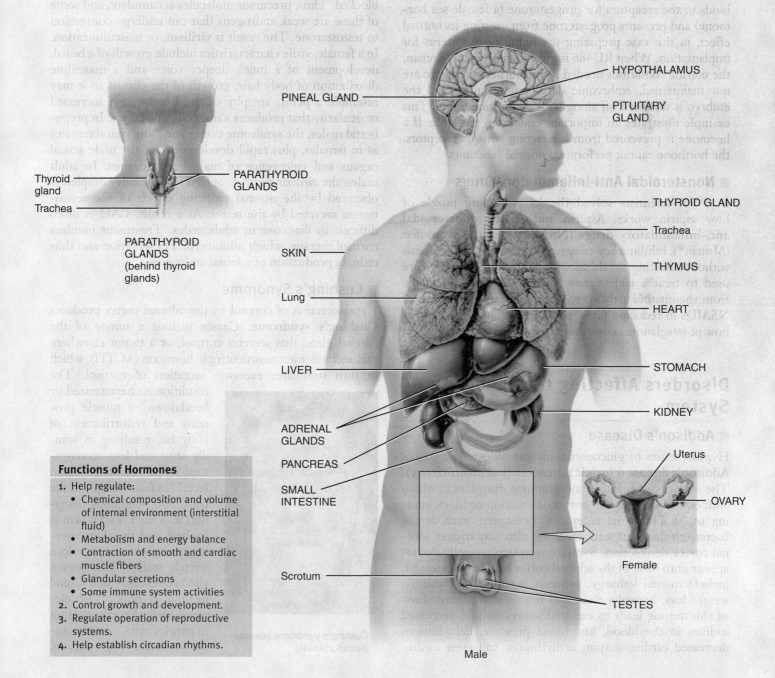

Thyroid gland

Trachea

PARATHYROID GLANDS

PARATHYROID GLANDS (behind thyroid glands)

HYPOTHALAMUS

PINEAL GLAND

PITUITARY GLAND

THYROID GLAND

Trachea

SKIN

THYMUS

Lung

HEART

LIVER

STOMACH

KIDNEY

ADRENAL GLANDS

Uterus

PANCREAS

SMALL INTESTINE

OVARY

Scrotum

Female

TESTES

Male

Functions of Hormones

1. Help regulate:
   - Chemical composition and volume of internal environment (interstitial fluid)
   - Metabolism and energy balance
   - Contraction of smooth and cardiac muscle fibers
   - Glandular secretions
   - Some immune system activities
2. Control growth and development.
3. Regulate operation of reproductive systems.
4. Help establish circadian rhythms.

Tests, Procedures, and Techniques

Administering Hormones

Both steroid hormones and thyroid hormones are effective when taken by mouth. They are not split apart during digestion and easily cross the intestinal lining because they are lipid-soluble. By contrast, peptide and protein hormones, such as insulin, are not effective oral medications because digestive enzymes destroy them by breaking their peptide bonds. This is why people who need insulin must take it by injection.

Blocking Hormone Receptors

Synthetic hormones that *block the receptors* for some naturally occurring hormones are available as drugs. For example, RU486 (mifepristone), which is used to induce abortion, binds to the receptors for progesterone (a female sex hormone) and prevents progesterone from exerting its normal effect, in this case preparing the lining of the uterus for implantation. When RU486 is given to a pregnant woman, the uterine conditions needed for nurturing an embryo are not maintained, embryonic development stops, and the embryo is sloughed off along with the uterine lining. This example illustrates an important endocrine principle: If a hormone is prevented from interacting with its receptors, the hormone cannot perform its normal functions.

Nonsteroidal Anti-inflammatory Drugs

In 1971, scientists solved the long-standing puzzle of how aspirin works. Aspirin and related **nonsteroidal anti-inflammatory drugs (NSAIDs)**, such as ibuprofen (Motrin®), inhibit a key enzyme in prostaglandin synthesis without affecting synthesis of leukotrienes. NSAIDs are used to treat a wide variety of inflammatory disorders, from rheumatoid arthritis to tennis elbow. The success of NSAIDs in reducing fever, pain, and inflammation shows how prostaglandins contribute to these woes. ■

Disorders Affecting the Endocrine System

Addison's Disease

Hyposecretion of glucocorticoids and aldosterone causes **Addison's disease (chronic adrenocortical insufficiency).** The majority of cases are autoimmune disorders in which antibodies cause adrenal cortex destruction or block binding of ACTH to its receptors. Pathogens, such as the bacterium that causes tuberculosis, also may trigger adrenal cortex destruction. Symptoms, which typically do not appear until 90% of the adrenal cortex has been destroyed, include mental lethargy, anorexia, nausea and vomiting, weight loss, hypoglycemia, and muscular weakness. Loss of aldosterone leads to elevated potassium and decreased sodium in the blood, low blood pressure, dehydration, decreased cardiac output, arrhythmias, and even cardiac arrest. The skin may have a "bronzed" appearance that often is mistaken for a suntan. Such was true in the case of President John F. Kennedy, whose Addison's disease was known to only a few while he was alive. Treatment consists of replacing glucocorticoids and mineralocorticoids and increasing sodium in the diet.

Congenital Adrenal Hyperplasia

Congenital adrenal hyperplasia (CAH) is a genetic disorder in which one or more enzymes needed for synthesis of cortisol are absent. Because the cortisol level is low, secretion of ACTH by the anterior pituitary is high due to lack of negative feedback inhibition. ACTH in turn stimulates growth and secretory activity of the adrenal cortex. As a result, both adrenal glands are enlarged. However, certain steps leading to synthesis of cortisol are blocked. Thus, precursor molecules accumulate, and some of these are weak androgens that can undergo conversion to testosterone. The result is **virilism,** or masculinization. In a female, virile characteristics include growth of a beard, development of a much deeper voice and a masculine distribution of body hair, growth of the clitoris so it may resemble a penis, atrophy of the breasts, and increased muscularity that produces a masculine physique. In prepubertal males, the syndrome causes the same characteristics as in females, plus rapid development of the male sexual organs and emergence of male sexual desires. In adult males, the virilizing effects of CAH are usually completely obscured by the normal virilizing effects of the testosterone secreted by the testes. As a result, CAH is often difficult to diagnose in adult males. Treatment involves cortisol therapy, which inhibits ACTH secretion and thus reduces production of adrenal androgens.

Cushing's Syndrome

Hypersecretion of cortisol by the adrenal cortex produces **Cushing's syndrome.** Causes include a tumor of the adrenal gland that secretes cortisol, or a tumor elsewhere that secretes adrenocorticotropic hormone (ACTH), which in turn stimulates excessive secretion of cortisol. The condition is characterized by breakdown of muscle proteins and redistribution of body fat, resulting in spindly arms and legs accompanied by a rounded "moon face," "buffalo hump" on the back, and pendulous (hanging) abdomen. Facial skin is flushed, and the skin covering the abdomen develops stretch marks. The person also bruises easily, and wound healing is poor. The elevated level of cortisol causes hyperglycemia, osteoporosis, weakness, hypertension,

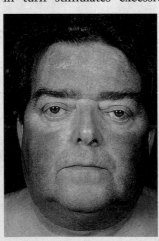

Cushing's syndrome (excess glucocorticoids)

increased susceptibility to infection, decreased resistance to stress, and mood swings. People who need long-term glucocorticoid therapy—for instance, to prevent rejection of a transplanted organ—may develop a cushinoid appearance.

■ Diabetes Insipidus

The most common abnormality associated with dysfunction of the posterior pituitary is **diabetes insipidus** (dī-a-BĒ-tēs in-SIP-i-dus; *diabetes* = overflow; *insipidus* = tasteless) or **DI**. This disorder is due to defects in antidiuretic hormone (ADH) receptors or an inability to secrete ADH. *Neurogenic diabetes insipidus* results from hyposecretion of ADH, usually caused by a brain tumor, head trauma, or brain surgery that damages the posterior pituitary or the hypothalamus. In *nephrogenic diabetes insipidus*, the kidneys do not respond to ADH. The ADH receptors may be nonfunctional, or the kidneys may be damaged. A common symptom of both forms of DI is excretion of large volumes of urine, with resulting dehydration and thirst. Bed-wetting is common in afflicted children. Because so much water is lost in the urine, a person with DI may die of dehydration if deprived of water for only a day or so.

Treatment of neurogenic diabetes insipidus involves hormone replacement, usually for life. Either subcutaneous injection or nasal spray application of ADH analogs is effective. Treatment of nephrogenic diabetes insipidus is more complex and depends on the nature of the kidney dysfunction. Restriction of salt in the diet and, paradoxically, the use of certain diuretic drugs, are helpful.

■ Diabetes Mellitus

The most common endocrine disorder is **diabetes mellitus** (MEL-i-tus; *melli-* = honey sweetened), caused by an inability to produce or use insulin. Diabetes mellitus is the fourth leading cause of death by disease in the United States, primarily because of its damage to the cardiovascular system. Because insulin is unavailable to aid transport of glucose into body cells, blood glucose level is high and glucose "spills" into the urine (glucosuria). Hallmarks of diabetes mellitus are the three "polys": *polyuria*, excessive urine production due to an inability of the kidneys to reabsorb water; *polydipsia*, excessive thirst; and *polyphagia*, excessive eating.

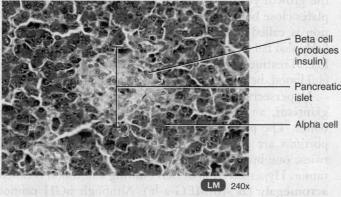

Beta cell (produces insulin)

Pancreatic islet

Alpha cell

LM 240x

Beta cells are the source of insulin

Both genetic and environmental factors contribute to onset of the two types of diabetes mellitus—type 1 and type 2—but the exact mechanisms are still unknown. In **type 1 diabetes** insulin level is low because the person's immune system destroys the pancreatic beta cells. It is also called **insulin-dependent diabetes mellitus (IDDM)** because insulin injections are required to prevent death. Most commonly, IDDM develops in people younger than age 20, though it persists throughout life. By the time symptoms of IDDM arise, 80–90% of the islet beta cells have been destroyed. IDDM is most common in northern Europe, especially in Finland where nearly 1% of the population develops IDDM by 15 years of age. In the United States, IDDM is 1.5–2.0 times more common in whites than in African American or Asian populations.

The cellular metabolism of an untreated type 1 diabetic is similar to that of a starving person. Because insulin is not present to aid the entry of glucose into body cells, most cells use fatty acids to produce ATP. Stores of triglycerides in adipose tissue are catabolized to yield fatty acids and glycerol. The byproducts of fatty acid breakdown—organic acids called ketones or ketone bodies—accumulate. Buildup of ketones causes blood pH to fall, a condition known as **ketoacidosis.** Unless treated quickly, ketoacidosis can cause death.

The breakdown of stored triglycerides also causes weight loss. As lipids are transported by the blood from storage depots to cells, lipid particles are deposited on the walls of blood vessels, leading to atherosclerosis and a multitude of cardiovascular problems, including cerebrovascular insufficiency, ischemic heart disease, peripheral vascular disease, and gangrene. A major complication of diabetes is loss of vision due either to cataracts (excessive glucose attaches to lens proteins, causing cloudiness) or to damage to blood vessels of the retina. Severe kidney problems also may result from damage to renal blood vessels.

Type 1 diabetes is treated through self-monitoring of blood glucose level (up to 7 times daily) (Figure 8.3b), regular meals containing 45–50% carbohydrates and less than 30% fats, exercise, and periodic insulin injections (up to 3 times a day) (Fig. 8.3c). Several implantable pumps are available to provide insulin without the need for repeated injections. Because they lack a reliable glucose sensor, however, the person must self-monitor blood glucose level to determine insulin doses. It is also possible to successfully transplant a pancreas, but immunosuppressive drugs must then be taken for life. Another promising approach under investigation is transplantation of isolated islets in semipermeable hollow tubes. The tubes allow glucose and insulin to enter and leave but prevent entry of immune system cells that might attack the islet cells.

Type 2 diabetes, also called **non-insulin-dependent diabetes mellitus (NIDDM),** is much more common than type 1, representing more than 90% of all cases. Type 2 diabetes most often occurs in obese people who are over age 35. However, the number of obese children and teenagers with type 2 diabetes is increasing. Clinical symptoms

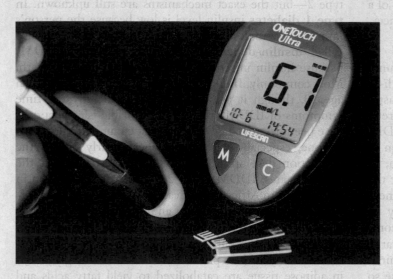

Self-monitoring blood glucose level

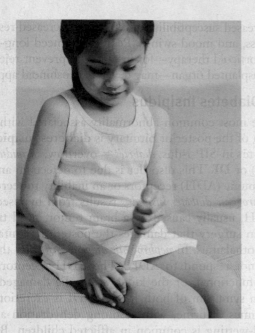

Insulin injection

are mild, and the high glucose levels in the blood often can be controlled by diet, exercise, and weight loss. Sometimes, a drug such as *glyburide* (DiaBeta®) is used to stimulate secretion of insulin by pancreatic beta cells. Although some type 2 diabetics need insulin, many have a sufficient amount (or even a surplus) of insulin in the blood. For these people, diabetes arises not from a shortage of insulin but because target cells become less sensitive to it due to down-regulation of insulin receptors.

Hyperinsulinism most often results when a diabetic injects too much insulin. The main symptom is **hypoglycemia,** decreased blood glucose level, which occurs because the excess insulin stimulates too much uptake of glucose by body cells. The resulting hypoglycemia stimulates the secretion of epinephrine, glucagon, and human growth hormone. As a consequence, anxiety, sweating, tremor, increased heart rate, hunger, and weakness occur. When blood glucose falls, brain cells are deprived of the steady supply of glucose they need to function effectively. Severe hypoglycemia leads to mental disorientation, convulsions, unconsciousness, and shock. Shock due to an insulin overdose is termed **insulin shock.** Death can occur quickly unless blood glucose level is raised. From a clinical standpoint, a diabetic suffering from either a hyperglycemia or a hypoglycemia crisis can have very similar symptoms—mental changes, coma, seizures, and so on. It is important to quickly and correctly identify the cause of the underlying symptoms and treat them appropriately.

■ Parathyroid Gland Disorders

Hypoparathyroidism—too little parathyroid hormone—leads to a deficiency of blood Ca^{2+}, which causes neurons and muscle fibers to depolarize and produce action poten-

tials spontaneously. This leads to twitches, spasms, and **tetany** (maintained contraction) of skeletal muscle. The leading cause of hypoparathyroidism is accidental damage to the parathyroid glands or to their blood supply during thyroidectomy surgery.

Hyperparathyroidism, an elevated level of parathyroid hormone, most often is due to a tumor of one of the parathyroid glands. An elevated level of PTH causes excessive resorption of bone matrix, raising the blood levels of calcium and phosphate ions and causing bones to become soft and easily fractured. High blood calcium level promotes formation of kidney stones. Fatigue, personality changes, and lethargy are also seen in patients with hyperparathyroidism.

■ Pituitary Dwarfism, Giantism, and Acromegaly

Several disorders of the anterior pituitary involve human growth hormone (hGH). Hyposecretion of hGH during the growth years slows bone growth, and the epiphyseal plates close before normal height is reached. This condition is called **pituitary dwarfism.** Other organs of the body also fail to grow, and the body proportions are childlike. Treatment requires administration of hGH during childhood, before the epiphyseal plates close.

Hypersecretion of hGH during childhood causes **giantism,** an abnormal increase in the length of long bones. The person grows to be very tall, but body proportions are about normal. Figure 8.4a shows identical twins; one brother developed giantism due to a pituitary tumor. Hypersecretion of hGH during adulthood is called **acromegaly** (ak'-rō-MEG-a-lē). Although hGH cannot produce further lengthening of the long bones because

A 22-year old man with pituitary giantism shown beside his identical twin

Acromegaly (excess hGH during adulthood)

the epiphyseal plates are already closed, the bones of the hands, feet, cheeks, and jaws thicken and other tissues enlarge. In addition, the eyelids, lips, tongue, and nose enlarge, and the skin thickens and develops furrows, especially on the forehead and soles.

Posttraumatic Stress Disorder

Posttraumatic stress disorder (PTSD) is an anxiety disorder that may develop in an individual who has experienced, witnessed, or learned about a physically or psychologically

distressing event. The immediate cause of PTSD appears to be the specific stressors associated with the events. Among the stressors are terrorism, hostage taking, imprisonment, military duty, serious accidents, torture, sexual or physical abuse, violent crimes, school shootings, massacres, and natural disasters. In the United States, PTSD affects 10% of females and 5% of males. Symptoms of PTSD include reexperiencing the event through nightmares or flashbacks; avoidance of any activity, person, place, or event associated with the stressors; loss of interest and lack of motivation; poor concentration; irritability; and insomnia. Treatment may include the use of antidepressants, mood stabilizers, and antianxiety and antipsychotic agents.

Seasonal Affective Disorder and Jet Lag

Seasonal affective disorder (SAD) is a type of depression that afflicts some people during the winter months, when day length is short. It is thought to be due, in part, to overproduction of melatonin. Full-spectrum bright-light therapy—repeated doses of several hours of exposure to artificial light as bright as sunlight—provides relief for some people. Three to six hours of exposure to bright light also appears to speed recovery from jet lag, the fatigue suffered by travelers who quickly cross several time zones.

Thyroid Gland Disorders

Thyroid gland disorders affect all major body systems and are among the most common endocrine disorders. **Congenital hypothyroidism,** hyposecretion of thyroid hormones that is present at birth, has devastating consequences if not treated promptly. Previously termed *cretinism,* this condition causes severe mental retardation and stunted bone growth. At birth, the baby typically is normal because lipid-soluble maternal thyroid hormones crossed the placenta during pregnancy and allowed normal development. Most states require testing of all newborns to ensure adequate thyroid function. If congenital hypothyroidism exists, oral thyroid hormone treatment must be started soon after birth and continued for life.

Hypothyroidism during the adult years produces **myxedema** (mix-e-DĒ-ma), which occurs about five times more often in females than in males. A hallmark of this disorder is edema (accumulation of interstitial fluid) that causes the facial tissues to swell and look puffy. A person with myxedema has a slow heart rate, low body temperature, sensitivity to cold, dry hair and skin, muscular weakness, general lethargy, and a tendency to gain weight easily. Because the brain has already reached maturity, mental retardation does not occur, but the person may be less alert. Oral thyroid hormones reduce the symptoms.

The most common form of hyperthyroidism is **Graves' disease,** which also occurs seven to ten times more often in females than in males, usually before age 40. Graves' disease is an autoimmune disorder in which the person produces antibodies that mimic the action of thyroid-stimulating hormone (TSH). The antibodies continually stimulate the thyroid gland to grow and produce thyroid

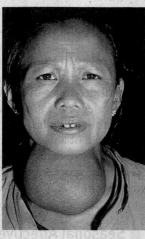

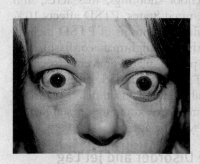

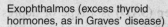

Exophthalmos (excess thyroid
hormones, as in Graves' disease)

Goiter (enlargement of
thyroid gland)

hormones. A primary sign is an enlarged thyroid, which may be two to three times its normal size. Graves' patients often have a peculiar edema behind the eyes, called **exophthalmos** (ek'-sof-THAL-mos), which causes the eyes to protrude. Treatment may include surgical removal of part or all of the thyroid gland (thyroidectomy), the use of radioactive iodine (^{131}I) to selectively destroy thyroid tissue, and the use of antithyroid drugs to block synthesis of thyroid hormones.

A **goiter** (GOY-ter; *guttur* = throat) is simply an enlarged thyroid gland. It may be associated with hyperthyroidism, hypothyroidism, or **euthyroidism** (*eu* = good), which means normal secretion of thyroid hormone. In some places in the world, dietary iodine intake is inadequate; the resultant low level of thyroid hormone in the blood stimulates secretion of TSH, which causes thyroid gland enlargement.

Additional Clinical Considerations

■ Diabetogenic Effect of hGH

One symptom of excess human growth hormone (hGH) is hyperglycemia. Persistent hyperglycemia in turn stimulates the pancreas to secrete insulin continually. Such excessive stimulation, if it lasts for weeks or months, may cause "beta-cell burnout," a greatly decreased capacity of pancreatic beta cells to synthesize and secrete insulin. Thus, excess secretion of human growth hormone may have a **diabetogenic effect**; that is, it causes diabetes mellitus (lack of insulin activity).

Oxytocin and Childbirth

Years before oxytocin was discovered, it was common practice in midwifery to let a first-born twin nurse at the mother's breast to speed the birth of the second child. Now we know why this practice is helpful—it stimulates the release of oxytocin. Even after a single birth, nursing promotes expulsion of the placenta (afterbirth) and helps the uterus regain its smaller size. Synthetic OT (Pitocin) often is given to induce labor or to increase uterine tone and control hemorrhage just after giving birth.

Pheochromocytomas

Usually benign tumors of the chromaffin cells of the adrenal medulla, called **pheochromocytomas** (fē-ō-krō'-mō-si-TŌ-mas; *pheo-* = dusky; *chromo-* = color; *cyto-* = cell), cause hypersecretion of epinephrine and norepinephrine. The result is a prolonged version of the fight-or-flight response: rapid heart rate, high blood pressure, high levels of glucose in blood and urine, an elevated basal metabolic rate (BMR), flushed face, nervousness, sweating, and decreased gastrointestinal motility. Treatment is surgical removal of the tumor. ■

Medical Terminology

Gynecomastia (gī-ne'-kō-MAS-tē-a; *gyneco-* = woman; *mast-* = breast) Excessive development of mammary glands in a male. Sometimes a tumor of the adrenal gland may secrete sufficient amounts of estrogen to cause the condition.

Hirsutism (HER-soo-tizm; *hirsut-* = shaggy) Presence of excessive body and facial hair in a male pattern, especially in women; may be due to excess androgen production due to tumors or drugs.

Thyroid crisis (storm) A severe state of hyperthyroidism that can be life threatening. It is characterized by high body temperature, rapidheart rate, high blood pressure, gastrointestinal symptoms (abdominal pain, vomiting, diarrhea), agitation, tremors, confusion, seizures, and possibly coma.

Virilizing adenoma (*aden* = gland; *oma* = tumor) Tumor of the adrenal gland that liberates excessive androgens, causing virilism (masculinization) in females. Occasionally, adrenal tumor cells liberate estrogens to the extent that a male patient develops gynecomastia. Such a tumor is called a **feminizing adenoma**.

Chapter 9 | The Cardiovascular System

The cardiovascular system consists of the blood, the heart, and blood vessels. **Blood** transports oxygen, carbon dioxide, nutrients, and hormones to and from your body's cells. It also helps regulate body pH and temperature, and provides protection against disease. The **heart** pumps blood through blood vessels to all body tissues. **Blood vessels** provide the structures for the flow of blood to and

from the heart, and the exchange of nutrients and wastes in tissues. In addition, they play an important role in adjusting the velocity and volume of blood flow. The branch of science concerned with the study of blood, blood forming tissues, and the disorders associated with them is **hematology.** The scientific study of the normal heart and the diseases associated with it is known as **cardiology.**

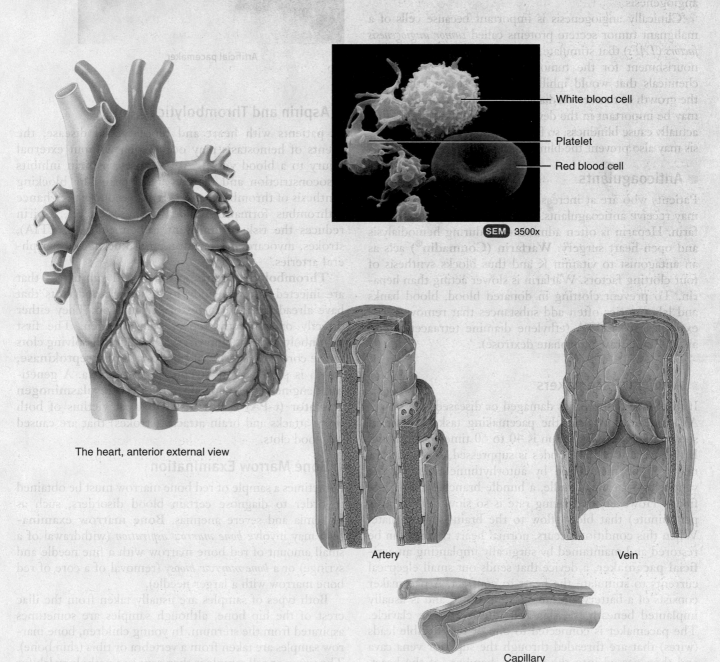

White blood cell

Platelet

Red blood cell

SEM 3500x

The heart, anterior external view

Artery

Vein

Capillary

Tests, Procedures, and Techniques

Angiogenesis and Disease

Angiogenesis (an'-jē-ō-JEN-e-sis; *angio-* = blood vessel; *-genesis* = production) refers to the growth of new blood vessels. It is an important process in embryonic and fetal development, and in postnatal life serves important functions such as wound healing, formation of a new uterine lining after menstruation, formation of the corpus luteum after ovulation, and development of blood vessels around obstructed arteries in the coronary circulation. Several proteins (peptides) are known to promote and inhibit angiogenesis.

Clinically angiogenesis is important because cells of a malignant tumor secrete proteins called *tumor angiogenesis factors* (*TAFs*) that stimulate blood vessel growth to provide nourishment for the tumor cells. Scientists are seeking chemicals that would inhibit angiogenesis and thus stop the growth of tumors. In diabetic retinopathy, angiogenesis may be important in the development of blood vessels that actually cause blindness, so finding inhibitors of angiogenesis may also prevent the blindness associated with diabetes.

Anticoagulants

Patients who are at increased risk of forming blood clots may receive anticoagulants. Examples are heparin or warfarin. Heparin is often administered during hemodialysis and open-heart surgery. **Warfarin (Coumadin®)** acts as an antagonist to vitamin K and thus blocks synthesis of four clotting factors. Warfarin is slower acting than heparin. To prevent clotting in donated blood, blood banks and laboratories often add substances that remove Ca^{2+}; examples are EDTA (ethylene diamine tetraacetic acid) and CPD (citrate phosphate dextrose).

Artificial Pacemakers

If the SA node becomes damaged or diseased, the slower AV node can pick up the pacemaking task. Its rate of spontaneous depolarization is 40 to 60 times per minute. If the activity of both nodes is suppressed, the heartbeat may still be maintained by autorhythmic fibers in the ventricles—the AV bundle, a bundle branch, or Purkinje fibers. However, the pacing rate is so slow (20–35 beats per minute) that blood flow to the brain is inadequate. When this condition occurs, normal heart rhythm can be restored and maintained by surgically implanting an **artificial pacemaker,** a device that sends out small electrical currents to stimulate the heart to contract. A pacemaker consists of a battery and impulse generator and is usually implanted beneath the skin just inferior to the clavicle. The pacemaker is connected to one or two flexible leads (wires) that are threaded through the superior vena cava and then passed into the various chambers of the heart. Many of the newer pacemakers, referred to as *activity-adjusted pacemakers*, automatically speed up the heartbeat during exercise.

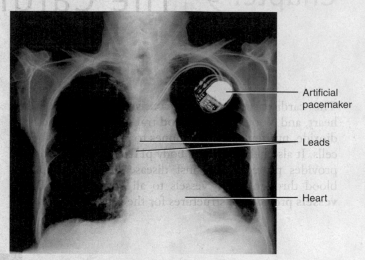

Artificial pacemaker

Labels: Artificial pacemaker, Leads, Heart

Aspirin and Thrombolytic Agents

In patients with heart and blood vessel disease, the events of hemostasis may occur even without external injury to a blood vessel. Atlow doses, **aspirin** inhibits vasoconstriction and platelet aggregation by blocking synthesis of thromboxane A2. It also reduces the chance ofthrombus formation. Due to these effects, aspirin reduces the risk of transient ischemic attacks (TIA), strokes, myocardial infarction, and blockage of peripheral arteries.

Thrombolytic agents are chemical substances that are injected into the body to dissolve blood clots that have already formed to restore circulation. They either directly or indirectly activate plasminogen. The first thrombolytic agent, approved in 1982 for dissolving clots in the coronary arteries of the heart, was **streptokinase,** which is produced by streptococcal bacteria. A genetically engineered version of human **tissue plasminogen activator (t-PA)** is now used to treat victims of both heart attacks and brain attacks (strokes) that are caused by blood clots.

Bone Marrow Examination

Sometimes a sample of red bone marrow must be obtained in order to diagnose certain blood disorders, such as leukemia and severe anemias. **Bone marrow examination** may involve *bone marrow aspiration* (withdrawal of a small amount of red bone marrow with a fine needle and syringe) or a *bone marrow biopsy* (removal of a core of red bone marrow with a larger needle).

Both types of samples are usually taken from the iliac crest of the hip bone, although samples are sometimes aspirated from the sternum. In young children, bone marrow samples are taken from a vertebra or tibia (shin bone). The tissue or cell sample is then sent to a pathology lab for analysis. Specifically, laboratory technicians look for signs of neoplastic (cancer) cells or other diseased cells to assist in diagnosis.

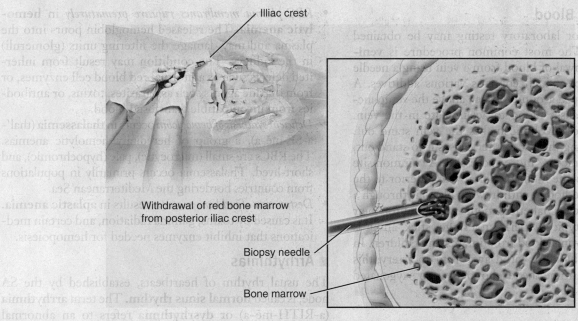

Illiac crest

Withdrawal of red bone marrow from posterior iliac crest

Biopsy needle

Bone marrow

Detailed view

Bone marrow examination

Cardiopulmonary Resuscitation

Because the heart lies between two rigid structures—the vertebral column and the sternum—external pressure on the chest (compression) can be used to force blood out of the heart and into the circulation. In cases in which the heart suddenly stops beating, **cardiopulmonary resuscitation (CPR)**—properly applied cardiac compressions, performed with artificial ventilation of the lungs via mouth-to-mouth respiration—saves lives. CPR keeps oxygenated blood circulating until the heart can be restarted.

In a 2007 Japanese study, researchers found that chest compressions alone are equally as effective as, if not better than, traditional CPR with lung ventilation. This is good news because it is easier for an emergency dispatcher to give instructions limited to chest compressions to frightened, nonmedical bystanders. As public fear of contracting contagious diseases such as hepatitis, HIV, and tuberculosis continues to rise, bystanders are much more likely to perform chest compressions alone than treatment involving mouth-to-mouth rescue breathing.

Complete Blood Count

A complete blood count (CBC) is a very valuable test that screens for anemia and various infections. Usually included are counts of RBCs, WBCs, and platelets per μL of whole blood; hematocrit; and differential white blood cell count. The amount of hemoglobin in grams per milliliter of blood also is determined. Normal hemoglobin ranges are: infants, 14–20 g/100 mL of blood; adult females, 12–16 g/100 mL of blood; and adult males, 13.5–18 g/100 mL of blood.

Medical Uses of Hemopoietic Growth Factors

Hemopoietic growth factors made available through recombinant DNA technology hold tremendous potential for medical uses when a person's natural ability to form new blood cells is diminished or defective. The artificial form of erythropoietin (Epoetin alfa) is very effective in treating the diminished red blood cell production that accompanies end-stage kidney disease. Granulocyte-macrophage colony-stimulating factor and granulocyte CSF are given to stimulate white blood cell formation in cancer patients who are undergoing chemotherapy, which kills red bone marrow cells as well as cancer cells because both cell types are undergoing mitosis. (Recall that white blood cells help protect against disease.) Thrombopoietin shows great promise for preventing the depletion of platelets, which are needed to help blood clot, during chemotherapy. CSFs and thrombopoietin also improve the outcome of patients who receive bone marrow transplants. Hemopoietic growth factors are also used to treat thrombocytopenia in neonates, other clotting disorders, and various types of anemia. Research on these medications is ongoing and shows a great deal of promise.

Reticulocyte Count

The rate of erythropoiesis is measured by a **reticulocyte count.** Normally, a little less than 1% of the oldest RBCs are replaced by newcomer reticulocytes on any given day. It then takes 1 to 2 days for the reticulocytes to lose the last vestiges of endoplasmic reticulum and become mature RBCs. Thus, reticulocytes account for about 0.5–1.5% of all RBCs in a normal blood sample. A low "retic" count in a person who is anemic might indicate a shortage of erythropoietin or an inability of the red bone marrow to respond to EPO, perhaps because of a nutritional deficiency or leukemia. A high "retic" count might indicate a good red bone marrow response to previous blood loss or to iron therapy in someone who had been iron deficient. It could also point to illegal use of Epoetin alfa by an athlete.

■ Withdrawing Blood

Blood samples for laboratory testing may be obtained in several ways. The most common procedure is **venipuncture,** withdrawal of blood from a vein using a needle and collecting tube, which contains various additives. A tourniquet is wrapped around the arm above the venipuncture site, which causes blood to accumulate in the vein. This increased blood volume makes the vein stand out. Opening and closing the fist further causes it to stand out, making the venipuncture more successful. A common site for venipuncture is the median cubital vein anterior to the elbow. Another method of withdrawing blood is through a **finger** or **heel stick.** Diabetic patients who monitor their daily blood sugar typically perform a finger stick, and it is often used for drawing blood from infants and children. In an **arterial stick,** blood is withdrawn from an artery; this test is used to determine the level of oxygen in oxygenated blood. ■

Disorders Affecting the Cardiovascular System

■ Anemia

Anemia is a condition in which the oxygen-carrying capacity of blood is reduced. All of the many types of anemia are characterized by reduced numbers of RBCs or a decreased amount of hemoglobin in the blood. The person feels fatigued and is intolerant of cold, both of which are related to lack of oxygen needed for ATP and heat production. Also, the skin appears pale, due to the low content of red-colored hemoglobin circulating in skin blood vessels. Among the most important causes and types of anemia are the following:

- *Inadequate absorption of iron, excessive loss of iron, increased iron requirement, or insufficient intake of iron* causes **iron-deficiency anemia,** the most common type of anemia. Women are at greater risk for iron-deficiency anemia due to menstrual blood losses and increased iron demands of the growing fetus during pregnancy. Gastrointestinal losses, such as those that occur with malignancy or ulceration, also contribute to this type of anemia.
- *Inadequate intake of vitamin B$_{12}$ or folic acid* causes **megaloblastic anemia** in which red bone marrow produces large, abnormal red blood cells (megaloblasts). It may also be caused by drugs that alter gastric secretion or are used to treat cancer.
- *Insufficient hemopoiesis* resulting from an inability of the stomach to produce intrinsic factor, which is needed for absorption of vitamin B$_{12}$ in the small intestine, causes **pernicious anemia.**
- *Excessive loss of RBCs* through bleeding resulting from large wounds, stomach ulcers, or especially heavy menstruation leads to **hemorrhagic anemia.**

- *RBC plasma membranes rupture prematurely* in **hemolytic anemia.** The released hemoglobin pours into the plasma and may damage the filtering units (glomeruli) in the kidneys. The condition may result from inherited defects such as abnormal red blood cell enzymes, or from outside agents such as parasites, toxins, or antibodies from incompatible transfused blood.
- *Deficient synthesis of hemoglobin* occurs in **thalassemia** (thal′-a-SĒ-mē-a), a group of hereditary hemolytic anemias. The RBCs are small (microcytic), pale (hypochromic), and short-lived. Thalassemia occurs primarily in populations from countries bordering the Mediterranean Sea.
- *Destruction of red bone marrow* results in **aplastic anemia.** It is caused by toxins, gamma radiation, and certain medications that inhibit enzymes needed for hemopoiesis.

■ Arrhythmias

The usual rhythm of heartbeats, established by the SA node, is called **normal sinus rhythm.** The term **arrhythmia** (a-RITH-mē-a) or **dysrhythmia** refers to an abnormal rhythm as a result of a defect in the conduction system of the heart. The heart may beat irregularly, too quickly, or too slowly. Symptoms include chest pain, shortness of breath, lightheadedness, dizziness, and fainting. Arrhythmias may be caused by factors that stimulate the heart such as stress, caffeine, alcohol, nicotine, cocaine, and certain drugs that contain caffeine or other stimulants. Arrhythmias may also be caused by a congenital defect, coronary artery disease, myocardial infarction, hypertension, defective heart valves, rheumatic heart disease, hyperthyroidism, and potassium deficiency.

Arrhythmias are categorized by their speed, rhythm, and origination of the problem. **Bradycardia** (brād-e-KAR-dē-a; *brady-* = slow) refers to a slow heart rate (below 50 beats per minute); **tachycardia** (tak-i-KAR-dē-a; *tachy-* = quick) refers to a rapid heart rate (over 100 beats per minute); and **fibrillation** (fi-bri-LĀ-shun) refers to rapid, uncoordinated heartbeats. Arrhythmias that begin in the atria are called **supraventricular** or **atrial arrhythmias;** those that originate in the ventricles are called **ventricular arrhythmias.**

- **Supraventricular tachycardia (SVT)** is a rapid but regular heart rate (160–200 beats) per minute that originates in the atria. The episodes begin and end suddenly and may last from a few minutes to many hours. SVTs can sometimes be stopped by maneuvers that stimulate the vagus (X) nerve and decrease heart rate. These include straining as if having a difficult bowel movement, rubbing the area over the carotid artery in the neck to stimulate the carotid sinus (not recommended for people over 50 since it may cause a stroke), and plunging the face into a bowl of ice-cold water. Treatment may also involve antiarrhythmic drugs and destruction of the abnormal pathway by radiofrequency ablation.
- **Heart block** is an arrhythmia that occurs when the electrical pathways between the atria and ventricles are

blocked, slowing the transmission of nerve impulses. The most common site of blockage is the atrioventricular node, a condition called *atrioventricular (AV) block*. In *first-degree AV block*, the P-Q interval is prolonged, usually because conduction through the AV node is slower than normal. In *second-degree AV block*, some of the action potentials from the SA node are not conducted through the AV node. The result is "dropped" beats because excitation doesn't always reach the ventricles. Consequently, there are fewer QRS complexes than P waves on the ECG. In *third-degree (complete) AV block*, no SA node action potentials get through the AV node. Autorhythmic fibers in the atria and ventricles pace the upper and lower chambers separately. With complete AV block, the ventricular contraction rate is less than 40 beats/min.

- **Atrial premature contraction (APC)** is a heartbeat that occurs earlier than expected and briefly interupts the normal heart rhythm. It often causes a sensation of a skipped heartbeat followed by a more forceful heartbeat. APCs originate in the atrial myocardium and are common in healthy individuals.
- **Atrial flutter** consists of rapid, regular atrial contractions (240–360 beats/min) accompanied by an atrioventricular (AV) block in which some of the nerve impulses from the SA node are not conducted through the AV node.

- **Atrial fibrillation (AF)** is a common arrhythmia, affecting mostly older adults, in which contraction of the atrial fibers is asynchronous (not in unison) so that atrial pumping ceases altogether. The atria may beat 300–600 beats/min. The ventricles may also speed up, resulting in a rapid heartbeat (up to 160 beats/min). The ECG of an individual with atrial fibrillation typically has no clearly defined P waves and irregularly spaced QRS complexes (and P-R intervals). Since the atria and ventricles do not beat in rhythm, the heartbeat is irregular in timing and strength. In an otherwise strong heart, atrial fibrillation reduces the pumping effectiveness of the heart by 20–30%. The most dangerous complication of atrial fibrillation is stroke since blood may stagnate in the atria and form blood clots. A stroke occurs when part of a blood clot occludes an artery supplying the brain.
- **Ventricular premature contraction.** Another form of arrhythmia arises when an *ectopic focus* (ek-TOP-ik), a region of the heart other than the conduction system, becomes more excitable than normal and causes an occasional abnormal action potential to occur. As a wave of depolarization spreads outward from the ectopic focus, it causes a **ventricular premature contraction (beat).** The contraction occurs early in diastole before the SA node is normally scheduled to discharge its

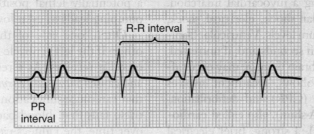

Normal electrocardiogram (ECG)

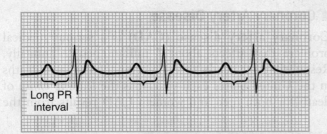

First-degree AV block

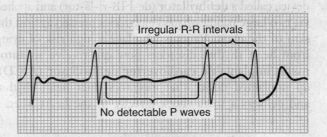

Atrial fibrillation

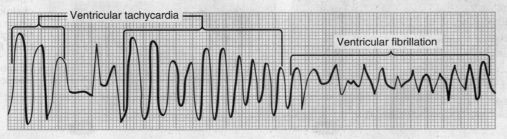

Ventricular tachycardia   Ventricular fibrillation action potential. Ventricular premature contractions may be relatively benign and may be caused by emotional stress, excessive intake of stimulants such as caffeine, alcohol, or nicotine, and lack of sleep. In other cases, the premature beats may reflect an underlying pathology.

- **Ventricular tachycardia (VT or V-tach)** is an arrhythmia that originates in the ventricles and is characterized by 4 or more ventricular premature contractions. It causes the ventricles to beat too fast (at least 120 beats/min). VT is almost always associated with heart disease or a recent myocardial infarction and may develop into a very serious arrhythmia called ventricular fibrillation (described shortly). Sustained VT is dangerous because the ventricles do not fill properly and thus do not pump sufficient blood. The result may be low blood pressure and heart failure.
- **Ventricular fibrillation (VF or V-fib)** is the most deadly arrhythmia, in which contractions of the ventricular fibers are completely asynchronous so that the ventricles quiver rather than contract in a coordinated way. As a result, ventricular pumping stops, blood ejection ceases, and circulatory failure and death occur unless there is immediate medical intervention. During ventricular fibrillation, the ECG has no clearly defined P waves, QRS complexes, or T waves. The most common cause of ventricular fibrillation is inadequate blood flow to the heart due to coronary artery disease, as occurs during a myocardial infarction. Other causes are cardiovascular shock, electrical shock, drowning, and very low potassium levels. Ventricular fibrillation causes unconsciousness in seconds and, if untreated, seizures occur and irreversible brain damage may occur after five minutes. Death soon follows. Treatment involves cardiopulmonary resuscitation (CPR) and defibrillation. In **defibrillation** (dē-fib-re-LĀ-shun), also called **cardioversion** (kar'-dē-ō-VER-shun), a strong, brief electrical current is passed to the heart and often can stop the ventricular fibrillation. The electrical shock is generated by a device called a **defibrillator** (de-FIB-ri-lā-tor) and applied via two large paddle-shaped electrodes pressed against the skin of the chest. Patients who face a high risk of dying from heart rhythm disorders now can receive an **automatic implantable cardioverter defibrillator (AICD),** an implanted device that monitors their heart rhythm and delivers a small shock directly to the heart when a

life-threatening rhythm disturbance occurs. Thousands of patients around the world have AICDs, including Dick Cheney, Vice President of the United States, who received a combination pacemaker-defibrillator in 2001. Also available are **automated external defibrillators (AEDs)** that function like AICDs, except that they are external devices. About the size of a laptop computer, AEDs are used by emergency response teams and are found increasingly in public places such as stadiums, casinos, airports, hotels, and shopping malls. Defibrillation may also be used as an emergency treatment for cardiac arrest.

■ Congestive Heart Failure

In **congestive heart failure (CHF),** there is a loss of pumping efficiency by the heart. Causes of CHF include coronary artery disease, congenital defects, long-term high blood pressure (which increases the afterload), myocardial infarctions (regions of dead heart tissue due to a previous heart attack), and valve disorders. As the pump becomes less effective, more blood remains in the ventricles at the end of each cycle, and gradually the end-diastolic volume (preload) increases. Initially, increased preload may promote increased force of contraction (the Frank–Starling law of the heart), but as the preload increases further, the heart is overstretched and contracts less forcefully. The result is a potentially lethal positive feedback loop: Less-effective pumping leads to even lower pumping capability.

Often, one side of the heart starts to fail before the other. If the left ventricle fails first, it can't pump out all the blood it receives. As a result, blood backs up in the lungs and causes *pulmonary edema*, fluid accumulation in the lungs that can cause suffocation if left untreated. If the right ventricle fails first, blood backs up in the systemic veins and, over time, the kidneys cause an increase in blood volume. In this case, the resulting *peripheral edema* usually is most noticeable in the feet and ankles.

■ Coronary Artery Disease

Coronary artery disease (CAD) is a serious medical problem that affects about 7 million people annually. Responsible for nearly three-quarters of a million deaths in the United States each year, it is the leading cause of death for both men and women. CAD results from the

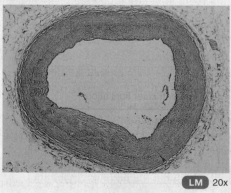

LM 20x

Normal artery

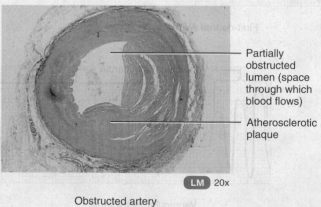

LM 20x

Partially obstructed lumen (space through which blood flows)

Atherosclerotic plaque

Obstructed artery

effects of the accumulation of atherosclerotic plaques (described shortly) in coronary arteries, which leads to a reduction in blood flow to the myocardium. Some individuals have no signs or symptoms; others experience angina pectoris (chest pain), and still others suffer heart attacks.

Risk Factors for CAD

People who possess combinations of certain risk factors are more likely to develop CAD. *Risk factors* (characteristics, symptoms, or signs present in a disease-free person that are statistically associated with a greater chance of developing a disease) include smoking, high blood pressure, diabetes, high cholesterol levels, obesity, "type A" personality, sedentary lifestyle, and a family history of CAD. Most of these can be modified by changing diet and other habits or can be controlled by taking medications. However, other risk factors are unmodifiable (beyond our control), including genetic predisposition (family history of CAD at an early age), age, and gender. For example, adult males are more likely than adult females to develop CAD; after age 70 the risks are roughly equal. Smoking is undoubtedly the number-one risk factor in all CAD-associated diseases, roughly doubling the risk of morbidity and mortality.

Development of Atherosclerotic Plaques

Although the following discussion applies to coronary arteries, the process can also occur in arteries outside the heart. Thickening of the walls of arteries and loss of elasticity are the main characteristics of a group of diseases called **arteriosclerosis** (ar-tē-rē-ō-skle-RŌ-sis; *sclero-* = hardening). One form of arteriosclerosis is **atherosclerosis** (ath-er-ō-skle-RŌ-sis), a progressive disease characterized by the formation in the walls of large and medium-sized arteries of lesions called **atherosclerotic plaques.**

To understand how atherosclerotic plaques develop, you will need to learn about the role of molecules produced by the liver and small intestine called **lipoproteins.** These spherical particles consist of an inner core of triglycerides and other lipids and an outer shell of proteins, phospholipids, and cholesterol. Like most lipids, cholesterol does not dissolve in water and must be made water-soluble in order to be transported in the blood. This is accomplished by combining it with lipoproteins. Two major lipoproteins are **low-density lipoproteins (LDLs)** and **high-density lipoproteins (HDLs).** LDLs transport cholesterol from the liver to body cells for use in cell membrane repair and the production of steroid hormones and bile salts. However, excessive amounts of LDLs promote atherosclerosis, so the cholesterol in these particles is commonly known as "bad cholesterol." HDLs, on the other hand, remove excess cholesterol from body cells and transport it to the liver for elimination. Because HDLs decrease blood cholesterol level, the cholesterol in HDLs is commonly referred to as "good cholesterol." Basically, you want your LDL concentration to be low and your HDL concentration to be high.

It has been learned recently that inflammation, a defensive response of the body to tissue damage, plays a key role in the development of atherosclerotic plaques. As a result of tissue damage, blood vessels dilate and increase their permeability, and phagocytes, including macrophages, appear in large numbers. The formation of atherosclerotic plaques begins when excess LDLs from the blood accumulate in the inner layer of an artery wall (layer closest to the bloodstream) and the lipids and proteins in the LDLs undergo oxidation and the proteins also bind to sugars. In response, endothelial and smooth muscle cells of the artery secrete substances that attract monocytes from the blood and convert them into macrophages. The macrophages then ingest and become so filled with oxidized LDL particles that they have a foamy appearance when viewed microscopically **(foam cells).** T cells (lymphocytes) follow monocytes into the inner lining of an artery where they release chemicals that intensify the inflammatory response. Together, the foam cells, macrophages, and T cells form a fatty streak, the beginning of an atherosclerotic plaque.

Macrophages secrete chemicals that cause smooth muscle cells of the middle layer of an artery to migrate to the top of the atherosclerotic plaque, forming a cap over it and thus walling it off from the blood.

Because most atherosclerotic plaques expand away from the bloodstream rather than into it, blood can still flow through the affected artery with relative ease, often for decades. Relatively few heart attacks occur when plaque in a coronary artery expands into the bloodstream and restricts blood flow. Most heart attacks occur when the cap over the plaque breaks open in response to chemicals produced by foam cells. In addition, T cells induce foam cells to produce tissue factor (TF), a chemical that begins the cascade of reactions that result in blood clot formation. If the clot in a coronary artery is large enough, it can significantly decrease or stop the flow of blood and result in a heart attack.

In recent years, a number of new risk factors (all modifiable) have been identified as significant predictors of CAD. **C-reactive proteins (CRPs)** are proteins produced by the liver or present in blood in an inactive form that are converted to an active form during inflammation. CRPs may play a direct role in the development of atherosclerosis by promoting the uptake of LDLs by macrophages. **Lipoprotein (a)** is an LDL-like particle that binds to endothelial cells, macrophages, and blood platelets, may promote the proliferation of smooth muscle fibers, and inhibits the breakdown of blood clots. **Fibrinogen** is a glycoprotein involved in blood clotting that may help regulate cellular proliferation, vasoconstriction, and platelet aggregation. **Homocysteine** is an amino acid that may induce blood vessel damage by promoting platelet aggregation and smooth muscle fiber proliferation.

■ Diagnosis of CAD

Many procedures may be employed to diagnose CAD; the specific procedure used will depend on the signs and symptoms of the individual.

A resting electrocardiogram is the standard test employed to diagnose CAD. **Stress testing** can also be performed. In an *exercise stress test*, the functioning of the heart is monitored when placed under physical stress by exercising using a treadmill, an exercise bicycle, or arm exercises. During the procedure, ECG recordings are monitored continuously and blood pressure is monitored at intervals. A *nonexercise (pharmacologic) stress test* is used for individuals who cannot exercise due to conditions such as arthritis. A medication is injected that stresses the heart to mimic the effects of exercise. During both exercise and nonexercise stress testing, **radionuclide imaging** may be performed to evaluate blood flow through heart muscle.

Diagnosis of CAD may also involve **echocardiography** (ek′-ō-kar-dē-OG-ra-fē), a technique that uses ultrasound waves to image the interior of the heart. Echocardiography allows the heart to be seen in motion and can be used to determine the size, shape, and functions of the chambers of the heart; the volume and velocity of blood pumped from the heart; the status of heart valves; the presence of birth defects; and abnormalities of the pericardium. A fairly recent technique for evaluating CAD is **electron beam computerized tomography (EBCT)**, which detects calcium deposits in coronary arteries. These calcium deposits are indicators of atherosclerosis.

Cardiac catheterization (kath′-e-ter-i-ZĀ-shun) is an invasive procedure used to visualize the heart's chambers, valves, and great vessels in order to diagnose and treat disease not related to abnormalities of the coronary arteries. It may also be used to measure pressure in the heart and great vessels; to assess cardiac output; to measure the flow of blood through the heart and great vessels; to identify the location of septal and valvular defects; and to take tissue and blood samples. The basic procedure involves inserting a long, flexible, radiopaque **catheter** (plastic tube) into a peripheral vein (for *right heart catheterization*) or a peripheral artery (for *left heart catheterization*) and guiding it under fluoroscopy (x-ray observation).

Coronary angiography (an′-jē-OG-ra-fē; *angio-* = blood vessel; *-grapho* = to write) is an invasive procedure used to obtain information about the coronary arteries. In the procedure, a catheter is inserted into an artery in the groin or wrist and threaded under fluoroscopy toward the heart and then into the coronary arteries. After the tip of the catheter is in place, a radiopaque contrast medium is injected into the coronary arteries. The radiographs of the arteries, called *angiograms*, appear in motion on a monitor and the information is recorded on a videotape or computer disc. Coronary angiography may be used to visualize coronary arteries and to inject clot-dissolving drugs, such as streptokinase or tissue plasminogen activa-tor (t-PA), into a coronary artery to dissolve an obstructing thrombus.

■ Treatment of CAD

Treatment options for CAD include **drugs** (antihypertensives, nitroglycerine, beta blockers, cholesterol-lowering drugs, and clot-dissolving agents) and various surgical and nonsurgical procedures designed to increase the blood supply to the heart.

Coronary artery bypass grafting (CABG) is a surgical procedure in which a blood vessel from another part of the body is attached ("grafted") to a coronary artery to bypass an area of blockage. A piece of the grafted blood vessel is sutured between the aorta and the unblocked portion of the coronary artery.

A nonsurgical procedure used to treat CAD is **percutaneous transluminal coronary angioplasty (PTCA)** (percutaneous = through the skin; *trans-* = across; *lumen* = an opening or channel in a tube; *angio-* = blood vessel; *-plasty* = to mold or to shape). In one variation of this procedure, a balloon catheter is inserted into an artery of an arm or leg and gently guided into a coronary artery. While dye is released, angiograms (x-rays of blood vessels) are taken to locate the plaques. Next, the catheter is advanced to the point of obstruction, and a balloonlike device is inflated with air to squash the plaque against the blood vessel wall. Because 30–50% of PTCA-opened arteries fail due to restenosis (renarrowing) within six months after the procedure is done, a stent may be inserted via a catheter. A **stent** is a metallic, fine wire tube that is permanently placed in an artery to keep the artery *patent* (open), permitting blood to circulate. Restenosis may be due to damage from the procedure itself, for PTCA may damage the arterial wall, leading to platelet activation, proliferation of smooth muscle fibers, and plaque formation. Recently, *drug-coated (drug-eluting) coronary stents* have been used to prevent restenosis. The stents are coated with one of several antiproliferative drugs (drugs that inhibit the proliferation of smooth muscle fibers of the middle layer of an artery) and anti-inflammatory drugs. It has been shown that drug-coated stents reduce the rate of restenosis when compared to bare-metal (noncoated) stents. In addition to balloon and stent angioplasty, laser-emitting catheters are used to vaporize plaques (excimer laser coronary angloplasy or ELCA) and small blades inside catheters are used to remove part of the plaque (directional coronary atherectomy).

One area of current research involves cooling the body's core temperature during procedures such as coronary artery bypass grafting (CABG). There have been some promising results from the application of cold therapy during a cerebral vascular accident (CVA or stroke). This research stemmed from observations of people who had suffered a hypothermic incident (such as cold-water drowning) and recovered with relatively minimal neurologic deficits.

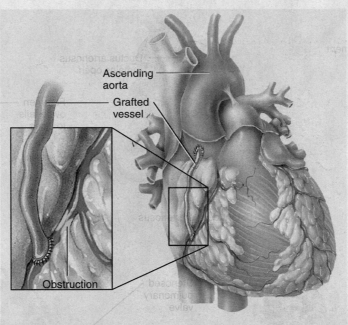

Ascending aorta

Grafted vessel

Obstruction

Coronary artery bypass grafting (CABG)

Balloon   Atherosclerotic   Narrowed lumen   Coronary
            plaque   of artery   artery

Balloon catheter with uninflated balloon is threaded
to obstructed area in artery

When balloon is inflated, it stretches arterial wall
and squashes atherosclerotic plaque

After lumen is widened, balloon is deflated and
catheter is withdrawn

Percutaneous transluminal coronary angioplasty (PTCA)

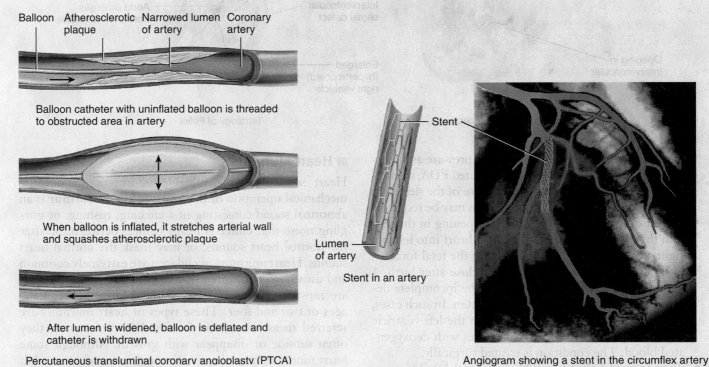

Stent

Lumen
of artery

Stent in an artery

Angiogram showing a stent in the circumflex artery

■ Congenital Heart Defects

A defect that is present at birth, and usually before, is called a **congenital defect.** Many such defects are not serious and may go unnoticed for a lifetime. Others are life threatening and must be surgically repaired. Among the several congenital defects that affect the heart are the following:

- **Coarctation** (kō′-ark-TĀ-shun) **of the aorta.** In this condition, a segment of the aorta is too narrow, and thus the flow of oxygenated blood to the body is reduced, the left ventricle is forced to pump harder, and high blood

pressure develops. Coarctation is usually repaired surgically by removing the area of obstruction. Surgical interventions that are done in childhood may require revisions in adulthood. Another surgical procedure is balloon dilation, insertion and inflation of a device in the aorta to stretch the vessel. A stent can be inserted and left in place to hold the vessel open.

- **Patent ductus arteriosus (PDA).** In some babies, the ductus arteriosus, a temporary blood vessel between the aorta and the pulmonary trunk, remains open rather than closing shortly after birth. As a result, aortic blood flows into the lower-pressure pulmonary trunk, thus

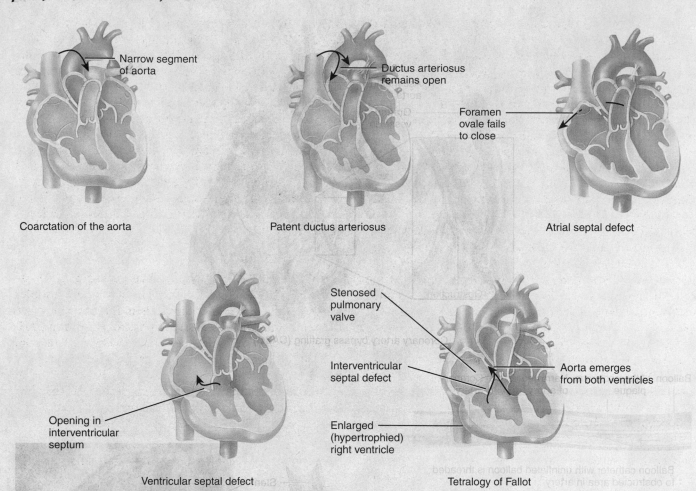

Coarctation of the aorta

Narrow segment of aorta

Patent ductus arteriosus

Ductus arteriosus remains open

Atrial septal defect

Foramen ovale fails to close

Ventricular septal defect

Opening in interventricular septum

Tetralogy of Fallot

Stenosed pulmonary valve

Interventricular septal defect

Enlarged (hypertrophied) right ventricle

Aorta emerges from both ventricles

increasing the pulmonary trunk blood pressure and over-working both ventricles. In uncomplicated PDA, medication can be used to facilitate the closure of the defect. In more severe cases, surgical intervention may be required.

- **Septal defect.** A septal defect is an opening in the septum that separates the interior of the heart into left and right sides. In an **atrial septal defect** the fetal foramen ovale between the two atria fails to close after birth. A **ventricular septal defect** is caused by incomplete development of the interventricular septum. In such cases, oxygenated blood flows directly from the left ventricle into the right ventricle, where it mixes with deoxygenated blood. The condition is treated surgically.

- **Tetralogy of Fallot** (tet-RAL-ō-jē of fal-Ō). This condition is a combination of four developmental defects: an interventricular septal defect, an aorta that emerges from both ventricles instead of from the left ventricle only, a stenosed pulmonary valve, and an enlarged right ventricle. There is a decreased flow of blood to the lungs and mixing of blood from both sides of the heart. This causes cyanosis, the bluish discoloration most easily seen in nail beds and mucous membranes when the level of deoxygenated hemoglobin is high; in infants, this condition is referred to as "blue baby." Despite the apparent complexity of this condition, surgical repair is usually successful.

■ Heart Murmurs

Heart sounds provide valuable information about the mechanical operation of the heart. A **heart murmur** is an abnormal sound consisting of a clicking, rushing, or gurgling noise that either is heard before, between, or after the normal heart sounds, or may mask the normal heart sounds. Heart murmurs in children are extremely common and usually do not represent a health condition. Murmurs are most frequently discovered in children between the ages of two and four. These types of heart murmurs are referred to as *innocent* or *functional heart murmurs;* they often subside or disappear with growth. Although some heart murmurs in adults are innocent, most often an adult murmur indicates a valve disorder. When a heart valve exhibits stenosis, the heart murmur is heard while the valve should be fully open but is not. For example, mitral stenosis produces a murmur during the relaxation period, between S2 and the next S1. An incompetent heart valve, by contrast, causes a murmur to appear when the valve should be fully closed but is not. So, a murmur due to mitral incompetence occurs during ventricular systole, between S1 and S2.

■ Heart Valve Disorders

When heart valves operate normally, they open fully and close completely at the proper times. A narrowing of a heart valve opening that restricts blood flow is known as **stenosis** (ste-NŌ-sis = a narrowing); failure of a valve to close completely is termed **insufficiency** or **incompetence.** In **mitral stenosis,** scar formation or a congenital defect causes narrowing of the mitral valve. One cause of **mitral insufficiency,** in which there is backflow of blood from the left ventricle into the left atrium, is **mitral valve prolapse (MVP).** In MVP one or both cusps of the mitral valve protrude into the left atrium during ventricular contraction. Mitral valve prolapse is one of the most common valvular disorders, affecting as much as 30% of the population. It is more prevalent in women than in men, and does not always pose a serious threat. In **aortic stenosis** the aortic valve is narrowed, and in **aortic insufficiency** there is backflow of blood from the aorta into the left ventricle.

Certain infectious diseases can damage or destroy the heart valves. One example is **rheumatic fever,** an acute systemic inflammatory disease that usually occurs after a streptococcal infection of the throat. The bacteria trigger an immune response in which antibodies produced to destroy the bacteria instead attack and inflame the connective tissues in joints, heart valves, and other organs. Even though rheumatic fever may weaken the entire heart wall, most often it damages the mitral and aortic valves.

Hemolytic Disease of the Newborn

The most common problem with Rh incompatibility, **hemolytic disease of the newborn (HDN),** may arise during pregnancy. Normally, no direct contact occurs between maternal and fetal blood while a woman is pregnant. However, if a small amount of Rh⁺ blood leaks from the fetus through the placenta into the bloodstream of an Rh⁻ mother, the mother will start to make anti-Rh antibodies. Because the greatest possibility of fetal blood leakage into the maternal circulation occurs at delivery, the firstborn baby usually is not affected. If the mother becomes pregnant again, however, her anti-Rh antibodies can cross the placenta and enter the bloodstream of the fetus. If the fetus is Rh⁻, there is no problem, because Rh⁻ blood does not have the Rh antigen. If the fetus is Rh⁺, however, agglutination and hemolysis brought on by fetal–maternal incompatibility may occur in the fetal blood.

An injection of anti-Rh antibodies called anti-Rh gamma globulin (RhoGAM®) can be given to prevent HDN. All Rh⁻ women should receive RhoGAM® soon after every delivery, miscarriage, or abortion. These antibodies bind to and inactivate the fetal Rh antigens before the mother's immune system can respond to the foreign antigens by producing her own anti-Rh antibodies.

Hemophilia

Hemophilia (hē-mō-FIL-ē-a; *-philia* = loving) is an inherited deficiency of clotting in which bleeding may occur spontaneously or after only minor trauma. It is the oldest known hereditary bleeding disorder; descriptions of the disease are found as early as the second century A.D. Hemophilia usually affects males and is sometimes

referred to as "the royal disease" because many descendants of Queen Victoria, beginning with one of her sons, were affected by the disease. Different types of hemophilia are due to deficiencies of different blood clotting factors and exhibit varying degrees of severity, ranging from mild to severe bleeding tendencies. Hemophilia is characterized by spontaneous or traumatic subcutaneous and intramuscular hemorrhaging, nosebleeds, blood in the urine, and hemorrhages in joints that produce pain and tissue damage. Treatment involves transfusions of fresh blood plasma or concentrates of the deficient clotting factor to relieve the tendency to bleed. Another treatment is the drug desmopressin (DDAVP), which can boost the levels of the clotting factors.

Hypertension

About 50 million Americans have **hypertension,** or persistently high blood pressure. It is the most common disorder affecting the heart and blood vessels and is the major cause of heart failure, kidney disease, and stroke. In May 2003, the Joint National Committee on Prevention, Detection, Evaluation, and Treatment of High Blood Pressure published new guidelines for hypertension because clinical studies have linked what were once considered fairly low blood pressure readings to an increased risk of cardiovascular disease. The new guidelines are as follows:

Category	Systolic (mmHg)	Diastolic (mmHg)
Normal	Less than 120 *and*	Less than 80
Prehypertension	120–139 *or*	80–89
Stage 1 hypertension	140–159 *or*	90–99
Stage 2 hypertension	Greater than 160 *or*	Greater than 100

Using the new guidelines, the normal classification was previously considered optimal; prehypertension now includes many more individuals previously classified as normal or high-normal; stage 1 hypertension is the same as in previous guidelines; and stage 2 hypertension now combines the previous stage 2 and stage 3 categories since treatment options are the same for the former stages 2 and 3.

Types and Causes of Hypertension

Between 90 and 95% of all cases of hypertension are **primary hypertension,** a persistently elevated blood pressure that cannot be attributed to any identifiable cause. The remaining 5–10% of cases are **secondary hypertension,** which has an identifiable underlying cause. Several disorders cause secondary hypertension:

- *Obstruction of renal blood flow* or disorders that damage renal tissue may cause the kidneys to release excessive amounts of renin into the blood. The resulting high level of angiotensin II causes vasoconstriction, thus increasing systemic vascular resistance.
- *Hypersecretion of aldosterone*—resulting, for instance, from a tumor of the adrenal cortex—stimulates excess

reabsorption of salt and water by the kidneys, which increases the volume of body fluids.

- *Hypersecretion of epinephrine and norepinephrine* by a **pheochromocytoma** (fē-ō-krō′-mō-sī-TO-ma), a tumor of the adrenal medulla. Epinephrine and norepinephrine increase heart rate and contractility and increase systemic vascular resistance.

Damaging Effects of Untreated Hypertension

High blood pressure is known as the "silent killer" because it can cause considerable damage to the blood vessels, heart, brain, and kidneys before it causes pain or other noticeable symptoms. It is a major risk factor for the number-one (heart disease) and number-three (stroke) causes of death in the United States. In blood vessels, hypertension causes thickening of the tunica media, accelerates development of atherosclerosis and coronary artery disease, and increases systemic vascular resistance. In the heart, hypertension increases the afterload, which forces the ventricles to work harder to eject blood.

The normal response to an increased workload due to vigorous and regular exercise is hypertrophy of the myocardium, especially in the wall of the left ventricle. An increased afterload, however, leads to myocardial hypertrophy that is accompanied by muscle damage and fibrosis (a buildup of collagen fibers between the muscle fibers). As a result, the left ventricle enlarges, weakens, and dilates. Because arteries in the brain are usually less protected by surrounding tissues than are the major arteries in other parts of the body, prolonged hypertension can eventually cause them to rupture, resulting in a stroke. Hypertension also damages kidney arterioles, causing them to thicken, which narrows the lumen; because the blood supply to the kidneys is thereby reduced, the kidneys secrete more renin, which elevates the blood pressure even more.

Lifestyle Changes to Reduce Hypertension

Although several categories of drugs (described next) can reduce elevated blood pressure, the following lifestyle changes are also effective in managing hypertension:

- *Lose weight.* This is the best treatment for high blood pressure short of using drugs. Loss of even a few pounds helps reduce blood pressure in overweight hypertensive individuals.
- *Limit alcohol intake.* Drinking in moderation may lower the risk of coronary heart disease, mainly among males over 45 and females over 55. Moderation is defined as no more than one 12-oz beer per day for females and no more than two 12-oz beers per day for males.
- *Exercise.* Becoming more physically fit by engaging in moderate activity (such as brisk walking) several times a week for 30 to 45 minutes can lower systolic blood pressure by about 10 mmHg.
- *Reduce intake of sodium (salt).* Roughly half the people with hypertension are "salt sensitive." For them, a high-salt diet appears to promote hypertension, and a low-salt diet can lower their blood pressure.

- *Maintain recommended dietary intake of potassium, calcium, and magnesium.* Higher levels of potassium, calcium, and magnesium in the diet are associated with a lower risk of hypertension.
- *Don't smoke.* Smoking has devastating effects on the heart and can augment the damaging effects of high blood pressure by promoting vasoconstriction.
- *Manage stress.* Various meditation and biofeedback techniques help some people reduce high blood pressure. These methods may work by decreasing the daily release of epinephrine and norepinephrine by the adrenal medulla.

Drug Treatment of Hypertension

Drugs having several different mechanisms of action are effective in lowering blood pressure. Many people are successfully treated with *diuretics*, agents that decrease blood pressure by decreasing blood volume because they increase elimination of water and salt in the urine. *ACE (angiotensin converting enzyme) inhibitors* block formation of angiotensin II and thereby promote vasodilation and decrease the secretion of aldosterone. *Beta blockers* reduce blood pressure by inhibiting the secretion of renin and by decreasing heart rate and contractility.

Vasodilators relax the smooth muscle in arterial walls, causing vasodilation and lowering blood pressure by lowering systemic vascular resistance. An important category of vasodilators are the *calcium channel blockers*, which slow the inflow of Ca^{2+} into vascular smooth muscle cells. They reduce the heart's workload by slowing Ca^{2+}

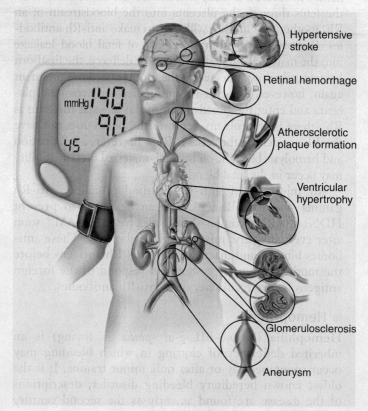

Hypertensive stroke

Retinal hemorrhage

Atherosclerotic plaque formation

Ventricular hypertrophy

Glomerulosclerosis

Aneurysm

Effects of Hypertension

entry into pacemaker cells and regular myocardial fibers, thereby decreasing heart rate and the force of myocardial contraction.

Leukemia

The term **leukemia** (loo-KĒ-mē-a; *leuko-* = white) refers to a group of red bone marrow cancers in which abnormal white blood cells multiply uncontrollably. The accumulation of the cancerous white blood cells in red bone marrow interferes with the production of red blood cells, white blood cells, and platelets. As a result the oxygen-carrying capacity of the blood is reduced, an individual is more susceptible to infection, and blood clotting is abnormal. In most leukemias, the cancerous white blood cells spread to the lymph nodes, liver, and spleen, causing them to enlarge. All leukemias produce the usual symptoms of anemia (fatigue, intolerance to cold, and pale skin). In addition, weight loss, fever, night sweats, excessive bleeding, and recurrent infections may occur.

In general, leukemias are classified as acute (symptoms develop rapidly) and chronic (symptoms may take years to develop). Adults may have either type, but children usually have the acute type.

The cause of most types of leukemia is unknown. However, certain risk factors have been implicated. These include exposure to radiation or chemotherapy for other cancers, genetics (some genetic disorders such as Down syndrome), environmental factors (smoking and benzene), and microbes such as the human T cell leukemia-lymphoma virus-1 (HTLV-1) and the Epstein-Barr virus.

Treatment options include chemotherapy, radiation, stem cell transplantation, interferon, antibodies, and blood transfusion.

Myocardial Ischemia and Infarction

Partial obstruction of blood flow in the coronary arteries may cause **myocardial ischemia** (is-KĒ-mē-a; *ische-* = to obstruct; *-emia* = in the blood), a condition of reduced blood flow to the myocardium. Usually, ischemia causes **hypoxia** (reduced oxygen supply), which may weaken cells without killing them. **Angina pectoris** (an-JĪ-na, or AN-ji-na, PEK-to-ris), which literally means "strangled chest," is a severe pain that usually accompanies myocardial ischemia. Typically, sufferers describe it as a tightness or squeezing sensation, as though the chest were in a vise. The pain associated with angina pectoris is often referred to the neck, chin, or down the left arm to the elbow. **Silent myocardial ischemia,** ischemic episodes without pain, is particularly dangerous because the person has no forewarning of an impending heart attack.

A complete obstruction to blood flow in a coronary artery may result in a **myocardial infarction** (in-FARK-shun), or **MI,** commonly called a *heart attack. Infarction* means the death of an area of tissue because of interrupted blood supply. Because the heart tissue distal to the obstruction dies and is replaced by noncontractile scar tissue, the heart muscle loses some of its strength. Depending on the size and location of the infarcted (dead) area, an infarction may disrupt the conduction system of the heart and cause sudden death by triggering ventricular fibrillation. Treatment for a myocardial infarction may involve injection of a thrombolytic (clot-dissolving) agent such as streptokinase or t-PA, plus heparin (an anticoagulant), or performing coronary angioplasty or coronary artery bypass grafting. Fortunately, heart muscle can remain alive in a resting person if it receives as little as 10–15% of its normal blood supply.

Myocarditis and Endocarditis

Myocarditis (mī-ō-kar-DĪ-tis) is an inflammation of the myocardium that usually occurs as a complication of a viral infection, rheumatic fever, or exposure to radiation or certain chemicals or medications. Myocarditis often has no symptoms. However, if they do occur, they may include fever, fatigue, vague chest pain, irregular or rapid heartbeat, joint pain, and breathlessness. Myocarditis is usually mild and recovery occurs within two weeks. Severe cases can lead to cardiac failure and death. Treatment consists of avoiding vigorous exercise, a low-salt diet, electrocardiographic monitoring, and treatment of the cardiac failure. **Endocarditis** (en′-dō-kar-DĪ-tis) refers to an inflammation of the endocardium and typically involves the heart valves. Most cases are caused by bacteria (bacterial endocarditis). Signs and symptoms of endocarditis include fever, heart murmur, irregular or rapid heartbeat, fatigue, loss of appetite, night sweats, and chills. Treatment is with intravenous antibiotics.

Pericarditis

Inflammation of the pericardium is called **pericarditis** (per′-i-kar-DĪ-tis). The most common type, *acute pericarditis,* begins suddenly and has no known cause in most cases but is sometimes linked to a viral infection. As a result of irritation to the pericardium, there is chest pain that may extend to the left shoulder and down the left arm (often mistaken for a heart attack) and pericardial friction rub (a scratchy or creaking sound heard through a stethoscope as the visceral layer of the serous pericardium rubs against the parietal layer of the serous pericardium). Acute pericarditis usually lasts for about one week and is treated with drugs that reduce inflammation and pain, such as ibuprofen or aspirin.

Chronic pericarditis begins gradually and is long-lasting. In one form of this condition, there is a buildup of pericardial fluid. If a great deal of fluid accumulates, this is a life-threatening condition because the fluid compresses the heart, a condition called *cardiac tamponade* (tam′-pon-ĀD). As a result of the compression, ventricular filling is decreased, cardiac output is reduced, venous return to the heart is diminished, blood pressure falls, and breathing is difficult. Most causes of chronic pericarditis involving cardiac tamponade are unknown, but it is sometimes caused by conditions such as cancer and tuberculosis. Treatment consists of draining the excess fluid through a needle passed into the pericardial cavity.

Sickle-Cell Disease

The RBCs of a person with **sickle-cell disease (SCD)** contain Hb-S, an abnormal kind of hemoglobin. When Hb-S gives up oxygen to the interstitial fluid, it forms long, stiff, rodlike structures that bend the erythrocyte into a sickle shape. The sickled cells rupture easily. Even though erythropoiesis is stimulated by the loss of the cells, it cannot keep pace with hemolysis. People with sickle-cell disease always have some degree of anemia and mild jaundice and may experience joint or bone pain, breathlessness, rapid heart rate, abdominal pain, fever, and fatigue as a result of tissue damage caused by prolonged recovery oxygen uptake (oxygen debt). Any activity that reduces the amount of oxygen in the blood, such as vigorous exercise, may produce a sickle-cell crisis (worsening of the anemia, pain in the abdomen and long bones of the limbs, fever, and shortness of breath).

Sickle-cell disease is inherited. People with two sickle-cell genes have severe anemia; those with only one defective gene have minor problems. Sickle-cell genes are found primarily among populations, or descendants of populations, that live

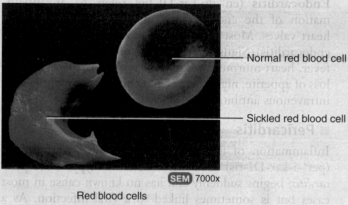

SEM 7000x
Red blood cells

Normal red blood cell

Sickled red blood cell

in the malaria belt around the world, including parts of Mediterranean Europe, sub-Saharan Africa, and tropical Asia. The gene responsible for the tendency of the RBCs to sickle also alters the permeability of the plasma membranes of sickled cells, causing potassium ions to leak out. Low levels of potassium kill the malaria parasites that may infect sickled cells. Because of this effect, a person with one normal gene and one sickle-cell gene has higher-than-average resistance to malaria. The possession of a single sickle-cell gene thus confers a survival advantage.

Treatment of SCD consists of administration of analgesics to relieve pain, fluid therapy to maintain hydration, oxygen to reduce the possibility of oxygen debt, antibiotics to counter infections, and blood transfusions. People who suffer from SCD have normal fetal hemoglobin (Hb-F), a slightly different form of hemoglobin that predominates at birth and is present in small amounts thereafter. In some patients with sickle-cell disease, a drug

called hydroxyurea promotes transcription of the normal Hb-F gene, elevates the level of Hb-F, and reduces the chance that the RBCs will sickle. Unfortunately, this drug also has toxic effects on the bone marrow; thus, its safety for long-term use is questionable.

Varicose Veins

Leaky venous valves can cause veins to become dilated and twisted in appearance, a condition called **varicose veins** or **varices** (VAR-i-sēz; varic- = a swollen vein). The singular is varix (VAR-iks). The condition may occur in the veins of almost any body part, but it is most common in the esophagus and in superficial veins of the lower limbs. Those in the lower limbs can range from cosmetic problems to serious medical conditions. The valvular defect may be congenital or may result from mechanical stress (prolonged standing or pregnancy) or aging. The leaking venous valves allow the backflow of blood from the deep veins to the less efficient superficial veins, where the blood pools. This creates pressure that distends the vein and allows fluid to leak into surrounding tissue. As a result, the affected vein and the tissue around it may become inflamed and painfully tender. Veins close to the surface of the legs, especially the saphenous vein, are highly susceptible to varicosities; deeper veins are not as vulnerable because surrounding skeletal muscles prevent their walls from stretching excessively. Varicose veins in the anal canal are referred to as *hemorrhoids*. Esophageal varices result from dilated veins in the walls of the lower part of the esophagus and sometimes the upper part of the stomach. Bleeding esophageal varices are life-threatening and are usually a result of chronic liver disease.

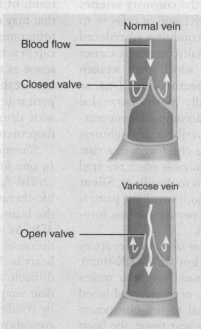

Normal vein

Blood flow

Closed valve

Varicose vein

Open valve

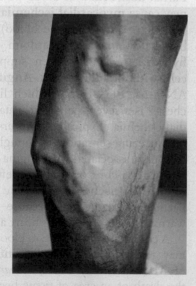

Dilated and twisted appearance of varicose veins in the leg

Several treatment options are available for varicose veins in the lower limbs. *Elastic stockings* (support hose) may be used for individuals with mild symptoms or for whom other options are not recommended. *Sclerotherapy* involves injection of a solution into varicose veins that damages the tunica interna by producing a harmless superficial thrombophlebitis (inflammation involving a blood clot). Healing of the damaged part leads to scar formation that occludes the vein. *Radiofrequency endovenous occlusion* involves the application of radiofrequency energy to heat up and close off varicose veins. *Laser occlusion* uses laser therapy to shut down veins. In a surgical procedure called *stripping*, veins may be removed. In this procedure, a flexible wire is threaded through the vein and then pulled out to strip (remove) it from the body.

Additional Clinical Considerations

Carotid Sinus Massage and Carotid Sinus Syncope

Because the carotid sinus is close to the anterior surface of the neck, it is possible to stimulate the baroreceptors there by putting pressure on the neck. Physicians sometimes use **carotid sinus massage,** which involves carefully massaging the neck over the carotid sinus, to slow heart rate in a person who has paroxysmal supraventricular tachycardia, a type of tachycardia that originates in the atria. Anything that stretches or puts pressure on the carotid sinus, such as hyperextension of the head, tight collars, or carrying heavy shoulder loads, may also slow heart rate and can cause **carotid sinus syncope,** fainting due to inappropriate stimulation of the carotid sinus baroreceptors.

Edema

If filtration greatly exceeds reabsorption, the result is **edema** (= swelling), an abnormal increase in interstitial fluid volume. Edema is not usually detectable in tissues until interstitial fluid volume has risen to 30% above normal. Edema can result from either excess filtration or inadequate reabsorption.

Two situations may cause excess filtration:

- *Increased capillary blood pressure* causes more fluid to be filtered from capillaries.
- *Increased permeability of capillaries* raises interstitial fluid osmotic pressure by allowing some plasma proteins to escape. Such leakiness may be caused by the destructive effects of chemical, bacterial, thermal, or mechanical agents on capillary walls.

One situation commonly causes inadequate reabsorption:

- *Decreased concentration of plasma proteins* lowers the blood colloid osmotic pressure. Inadequate synthesis or dietary intake or loss of plasma proteins is associated with liver disease, burns, malnutrition (for example, kwashiorkor), and kidney disease.

Iron Overload and Tissue Damage

Because free iron ions (Fe^{2+} and Fe^{3+}) bind to and damage molecules in cells or in the blood, transferrin and ferritin act as protective "protein escorts" during transport and storage of iron ions. As a result, plasma contains virtually no free iron. Furthermore, only small amounts are available inside body cells for use in synthesis of iron-containing molecules such as the cytochrome pigments needed for ATP production in mitochondria. In cases of **iron overload,** the amount of iron present in the body builds up. Because we have no method for eliminating excess iron, any condition that increases dietary iron absorption can cause iron overload. At some point, the proteins transferrin and ferritin become saturated with iron ions, and free iron level rises. Common consequences of iron overload are diseases of the liver, heart, pancreatic islets, and gonads. Iron overload also allows certain iron-dependent microbes to flourish. Such microbes normally are not pathogenic, but they multiply rapidly and can cause lethal effects in a short time when free iron is present.

Regeneration of Heart Cells

The heart of a heart attack survivor often has regions of infarcted (dead) cardiac muscle tissue that typically are replaced with noncontractile fibrous scar tissue over time. Our inability to repair damage from a heart attack has been attributed to a lack of stem cells in cardiac muscle and to the absence of mitosis in mature cardiac muscle fibers. A recent study of heart transplant recipients by American and Italian scientists, however, provides evidence for significant replacement of heart cells. The researchers studied men who had received a heart from a female, and then looked for the presence of a Y chromosome in heart cells. (All female cells except gametes have two X chromosomes and lack the Y chromosome.) Several years after the transplant surgery, between 7% and 16% of the heart cells in the transplanted tissue, including cardiac muscle fibers and endothelial cells in coronary arterioles and capillaries, had been replaced by the recipient's own cells, as evidenced by the presence of a Y chromosome. The study also revealed cells with some of the characteristics of stem cells in both transplanted hearts and control hearts. Evidently, stem cells can migrate from the blood into the heart and differentiate into functional muscle and endothelial cells. The hope is that researchers can learn how to "turn on" such regeneration of heart cells to treat people with heart failure or cardiomyopathy (diseased heart).

Syncope

Syncope (SIN-kō-pē), or fainting, is a sudden, temporary loss of consciousness that is not due to head trauma, followed by spontaneous recovery. It is most commonly due to cerebral ischemia, lack of sufficient blood flow to the brain. Syncope may occur for several reasons:

- *Vasodepressor syncope* is due to sudden emotional stress or real, threatened, or fantasized injury.

- *Situational syncope* is caused by pressure stress associated with urination, defecation, or severe coughing.
- *Drug-induced syncope* may be caused by drugs such as antihypertensives, diuretics, vasodilators, and tranquilizers.
- *Orthostatic hypotension*, an excessive decrease in blood pressure that occurs upon standing up, may cause fainting.

Medical Terminology

Acute normovolemic hemodilution (nor-mō-vō-LĒ-mik hē-mō-di-LOO-shun) Removal of blood immediately before surgery and its replacement with a cell-free solution to maintain sufficient blood volume for adequate circulation. At the end of surgery, once bleeding has been controlled, the collected blood is returned to the body.

Aneurysm (AN-Ū-rizm) A thin, weakened section of the wall of an artery or a vein that bulges outward, forming a balloonlike sac. Common causes are atherosclerosis, syphilis, congenital blood vessel defects, and trauma. If untreated, the aneurysm enlarges and the blood vessel wall becomes so thin that it bursts. The result is massive hemorrhage with shock, severe pain, stroke, or death. Treatment may involve surgery in which the weakened area of the blood vessel is removed and replaced with a graft of synthetic material.

Aortography (ā'-or-TOG-ra-fē) X-ray examination of the aorta and its main branches after injection of a radiopaque dye.

Asystole (ā-SIS-tō-lē; *a-* = without) Failure of the myocardium to contract.

Autologous preoperative transfusion (aw-TOL-o-gus trans-FŪ-zhun; *auto-* = self) Donating one's own blood; can be done up to 6 weeks before elective surgery. Also called **predonation**. This procedure eliminates the risk of incompatibility and blood-borne disease.

Blood bank A facility that collects and stores a supply of blood for future use by the donor or others. Because blood banks have additional and diverse functions (immunohematology reference work, continuing medical education, bone and tissue storage, and clinical consultation), they are more appropriately referred to as **centers of transfusion medicine.**

Cardiac arrest (KAR-dē-ak a-REST) A clinical term meaning cessation of an effective heartbeat. The heart may be completely stopped or in ventricular fibrillation.

Cardiomegaly (kar'-dē-ō-MEG-a-lē; *mega* = large) Heart enlargement.

Cardiac rehabilitation (rē-ha-bil-i-TĀ-shun) A supervised program of progressive exercise, psychological support, education, and training to enable a patient to resume normal activities following a myocardial infarction.

Cardiomyopathy (kar'-dē-ō-mī-OP-a-thē; *myo-* = muscle; *-pathos* = disease) A progressive disorder in which ventricular structure or function is impaired. In dilated car-

diomyopathy, the ventricles enlarge (stretch) and become weaker and reduce the heart's pumping action. In hypertrophic cardiomyopathy, the ventricular walls thicken and the pumping efficiency of the ventricles is reduced.

Carotid endarterectomy (ka-ROT-id end'-ar-ter-EK-tō-mē) The removal of atherosclerotic plaque from the carotid artery to restore greater blood flow to the brain.

Claudication (klaw'-di-KĀ-shun) Pain and lameness or limping caused by defective circulation of the blood in the vessels of the limbs.

Commotio cordis (kō-MŌ-shē-ō KOR-dis; *commotio-* = disturbance; *cordis* = heart) Damage to the heart, frequently fatal, as a result of a sharp, nonpenetrating blow to the chest while the ventricles are repolarizing.

Cor pulmonale (CP) (kor pul-mōn-ALE; *cor-* = heart; *pulmon-* = lung) A term referring to right ventricular hypertrophy from disorders that bring about hypertension (high blood pressure) in the pulmonary circulation.

Cyanosis (sī-a-NŌ-sis; *cyano-* = blue) Slightly bluish/dark-purple skin discoloration, most easily seen in the nail beds and mucous membranes, due to an increased quantity of reduced hemoglobin (hemoglobin not combined with oxygen) in systemic blood.

Deep vein thrombosis The presence of a thrombus (blood clot) in a deep vein of the lower limbs. It may lead to (1) pulmonary embolism, if the thrombus dislodges and then lodges within the pulmonary arterial blood flow, and (2) postphlebitic syndrome, which consists of edema, pain, and skin changes due to destruction of venous valves.

Doppler ultrasound scanning Imaging technique commonly used to measure blood flow. A transducer is placed on the skin and an image is displayed on a monitor that provides the exact position and severity of a blockage.

Ejection fraction The fraction of the end-diastolic volume (EDV) that is ejected during an average heartbeat. Equal to stroke volume (SV) divided by EDV.

Electrophysiological testing (e-lek'-trō-fiz'-ē-OL-ō-ji-kal) A procedure in which a catheter with an electrode is passed through blood vessels and introduced into the heart. It is used to detect the exact locations of abnormal electrical conduction pathways. Once an abnormal pathway is located, it can be destroyed by sending a current through the electrode, a procedure called *radiofrequency ablation*.

Femoral angiography An imaging technique in which a contrast medium is injected into the femoral artery and spreads to other arteries in the lower limb, and then a series of radiographs are taken of one or more sites. It is used to diagnose narrowing or blockage of arteries in the lower limbs.

Gamma globulin (GLOB-ū-lin) Solution of immunoglobulins from blood consisting of antibodies that react with specific pathogens, such as viruses. It is prepared by injecting the specific virus into animals, removing blood from the animals after antibodies have accumulated, isolating the antibodies, and injecting them into a human to provide short-term immunity.

Hemochromatosis (hē-mō-krō-ma-TŌ-sis; *chroma* = color) Disorder of iron metabolism characterized by excessive absorption of ingested iron and excess deposits of iron in tissues (especially the liver, heart, pituitary gland, gonads, and pancreas) that result in bronze discoloration of the skin, cirrhosis, diabetes mellitus, and bone and joint abnormalities.

Hemorrhage (HEM-or-ij; *rhegnynai* = bursting forth) Loss of a large amount of blood; can be either internal (from blood vessels into tissues) or external (from blood vessels directly to the surface of the body).

Hypotension (hī-pō-TEN-shun) Low blood pressure; most commonly used to describe an acute drop in blood pressure, as occurs during excessive blood loss.

Jaundice (*jaund-* = yellow) An abnormal yellowish discoloration of the sclerae of the eyes, skin, and mucous membranes due to excess bilirubin (yellow-orange pigment) in the blood. The three main categories of jaundice are *prehepatic jaundice*, due to excess production of bilirubin; *hepatic jaundice*, abnormal bilirubin processing by the liver caused by congenital liver disease, cirrhosis (scar tissue formation) of the liver, or hepatitis (liver inflammation); and *extrahepatic jaundice*, due to blockage of bile drainage by gallstones or cancer of the bowel or pancreas.

Normotensive (nor'-mō-TEN-siv) Characterized by normal blood pressure.

Occlusion (ō-KLOO-shun) The closure or obstruction of the lumen of a structure such as a blood vessel. An example is an atherosclerotic plaque in an artery.

Orthostatic hypotension (or'-thō-STAT-ik; *ortho-* = straight; *-static* = causing to stand) An excessive lowering of systemic blood pressure when a person assumes an erect or semi-erect posture; it is usually a sign of a disease. May be caused by excessive fluid loss, certain drugs, and cardiovascular or neurogenic factors. Also called **postural hypotension.**

Palpitation (pal'-pi-TĀ-shun) A fluttering of the heart or an abnormal rate or rhythm of the heart about which an individual is aware.

Paroxysmal tachycardia (par'-ok-SIZ-mal tak'-e-KAR-dē-a; *tachy-* = quick) A period of rapid heartbeats that begins and ends suddenly.

Phlebitis (fle-BĪ-tis; *phleb-* = vein) Inflammation of a vein, often in a leg.

Phlebotomist (fle-BOT-ō-mist; *phlebo-* = vein; *-tom* = cut) A technician who specializes in withdrawing blood.

Septicemia (sep'-ti-SĒ-mē-a; *septic-* = decay; *-emia* = condition of blood) Toxins or disease-causing bacteria in the blood. Also called "blood poisoning."

Sick sinus syndrome An abnormally functioning SA node that initiates heartbeats too slowly or rapidly, pauses too long between heartbeats, or stops producing heartbeats. Symptoms include lightheadedness, shortness of breath, loss of consciousness, and palpitations. It is caused by degeneration of cells in the SA node and is common in elderly persons. It is also related to coronary artery disease. Treatment consists of drugs to speed up or slow down the heart or implantation of an artificial pacemaker.

Sudden cardiac death The unexpected cessation of circulation and breathing due to an underlying heart disease such as ischemia, myocardial infarction, or a disturbance in cardiac rhythm.

Thrombectomy (throm-BEK-tō-mē; *thrombo-* = clot) An operation to remove a blood clot from a blood vessel.

Thrombocytopenia (throm'-bō-sī-tō-PĒ-nē-a; *-penia* = poverty) Very low platelet count that results in a tendency to bleed from capillaries.

Thrombophlebitis (throm'-bō-fle-BĪ-tis) Inflammation of a vein involving clot formation. Superficial thrombophlebitis occurs in veins under the skin, especially in the calf.

Venipuncture (VEN-i-punk-chur; *vena-* = vein) The puncture of a vein, usually to withdraw blood for analysis or introduce a solution, for example, an antibiotic. The median cubital vein is frequently used.

Venesection (vē'-ne-SEK-shun; *ven-* = vein) Opening of a vein for withdrawal of blood. Although **phlebotomy** (fle-BOT-ō-mē) is a synonym for venesection, in clinical practice phlebotomy refers to therapeutic bloodletting, such as the removal of some blood to lower its viscosity in a patient with polycythemia.

White coat (office) hypertension A stress-induced syndrome found in patients who have elevated blood pressure when being examined by health-care personnel, but otherwise have normal blood pressure.

Whole blood Blood containing all formed elements, plasma, and plasma solutes in natural concentrations.

Chapter 10

The Lymphatic System and Immunity

Immunity is the ability to ward off damage or disease through our defenses. *Innate immunity* refers to defenses that are present at birth. *Adaptive* [specific] *immunity* refers to defenses that involve specific recognition of a microbe once it has breached the innate immunity defenses. The body system responsible for adaptive immunity [and some aspects of innate immunity) is the **lymphatic system.** It is closely allied with the cardiovascular system. Adaptive immunity involves lymphocytes [a type of white blood cell] called T lymphocytes (T cells) and B lymphocytes (B cells), which protect all body systems from attack by harmful pathogens, foreign cells, and cancer cells.

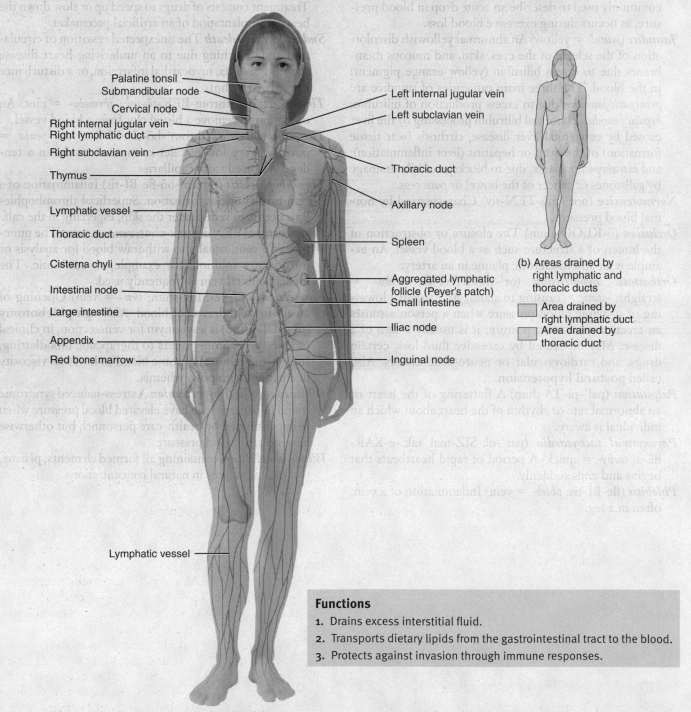

Palatine tonsil
Submandibular node
Cervical node
Right internal jugular vein
Right lymphatic duct
Right subclavian vein
Thymus
Lymphatic vessel
Thoracic duct
Cisterna chyli
Intestinal node
Large intestine
Appendix
Red bone marrow

Left internal jugular vein
Left subclavian vein
Thoracic duct
Axillary node
Spleen
Aggregated lymphatic follicle (Peyer's patch)
Small intestine
Iliac node
Inguinal node

Lymphatic vessel

(b) Areas drained by right lymphatic and thoracic ducts

Area drained by right lymphatic duct
Area drained by thoracic duct

Anterior view of principal components of lymphatic system

Functions

1. Drains excess interstitial fluid.
2. Transports dietary lipids from the gastrointestinal tract to the blood.
3. Protects against invasion through immune responses.

Tests, Procedures, and Techniques

■ Cytokine Therapy

Cytokine therapy is the use of cytokines to treat medical conditions. Interferons were the first cytokines shown to have limited effects against some human cancers. Alpha-interferon (Intron A®) is approved in the United States for treating Kaposi (KAP-ō-sē) sarcoma, a cancer that often occurs in patients infected with HIV, the virus that causes AIDS. Other approved uses for alpha-interferon include treating genital herpes caused by the herpes virus; treating hepatitis B and C, caused by the hepatitis B and C viruses; and treating hairy cell leukemia. A form of beta-interferon (Betaseron®) slows the progression of multiple sclerosis (MS) and lessens the frequency and severity of MS attacks. Of the interleukins, the one most widely used to fight cancer is interleukin-2. Although this treatment is effective in causing tumor regression in some patients, it also can be very toxic. Among the adverse effects are high fever, severe weakness, difficulty breathing due to pulmonary edema, and hypotension leading to shock.

■ Graft Rejection and Tissue Typing

Organ transplantation involves the replacement of an injured or diseased organ, such as the heart, liver, kidney, lungs, or pancreas, with an organ donated by another individual. Usually, the immune system recognizes the proteins in the transplanted organ as foreign and mounts both cell-mediated and antibody-mediated immune responses against them. This phenomenon is known as **graft rejection.** The success of an organ or tissue transplant depends on **histocompatibility** (his′-tō-kom-pat-i-BIL-i-tē)—that is, the tissue compatibility between the donor and the recipient. The more similar the MHC antigens, the greater the histocompatibility, and thus the greater the probability that the transplant will not be rejected. **Tissue typing (histocompatibility testing)** is done before any organ transplant. In the United States, a nationwide computerized registry helps physicians select the most histocompatible and needy organ transplant recipients whenever donor organs become available. The closer the match between the major histocompatibility complex proteins of the donor and recipient, the weaker is the graft rejection response.

To reduce the risk of graft rejection, organ transplant recipients receive immunosuppressive drugs. One such drug is *cyclosporine*, derived from a fungus, which inhibits secretion of interleukin-2 by helper T cells but has only a minimal effect on B cells. Thus, the risk of rejection is diminished while resistance to some diseases is maintained.

Disorders Affecting the Lymphatic System and Immunity

■ AIDS: Acquired Immunodeficiency Syndrome

Acquired immunodeficiency syndrome (AIDS) is a condition in which a person experiences a telltale assortment of infections due to the progressive destruction of immune system cells by the **human immunodeficiency virus (HIV).** AIDS represents the end stage of infection by HIV. A person who is infected with HIV may be symptom-free for many years, even while the virus is actively attacking the immune system. In the two decades after the first five cases were reported in 1981, 22 million people died of AIDS. Worldwide, 35 to 40 million people are currently infected with HIV.

■ HIV Transmission

Because HIV is present in the blood and some body fluids, it is most effectively transmitted (spread from one person to another) by actions or practices that involve the exchange of blood or body fluids between people. HIV is transmitted in semen or vaginal fluid during unprotected (without a condom) anal, vaginal, or oral sex. HIV also is transmitted by direct blood-to-blood contact, such as occurs among intravenous drug users who share hypodermic needles or health-care professionals who may be accidentally stuck by HIV-contaminated hypodermic needles. In addition, HIV can be transmitted from an HIV-infected mother to her baby at birth or during breast-feeding.

The chance of transmitting or of being infected by HIV during vaginal or anal intercourse can be greatly reduced—although not entirely eliminated—by the use of latex condoms. Public health programs aimed at encouraging drug users not to share needles have proven effective at checking the increase in new HIV infections in this population. Also, giving certain drugs to pregnant HIV-infected women greatly reduces the risk of transmission of the virus to their babies.

HIV is a very fragile virus; it cannot survive for long outside the human body. The virus is not transmitted by insect bites. One cannot become infected by casual physical contact with an HIV-infected person, such as by hugging or sharing household items. The virus can be eliminated from personal care items and medical equipment by exposing them to heat (135°F for 10 minutes) or by cleaning them with common disinfectants such as hydrogen peroxide, rubbing alcohol, household bleach, or germicidal cleansers such as Betadine or Hibiclens. Standard dishwashing and clothes washing also kill HIV.

■ HIV: Structure and Infection

HIV consists of an inner core of ribonucleic acid (RNA) covered by a protein coat (capsid). HIV is classified as a **retrovirus** since its genetic information is carried in RNA instead of DNA. Surrounding the HIV capsid is an envelope composed of a lipid bilayer that is penetrated by glycoproteins.

Outside a living host cell, a virus is unable to replicate. However, when a virus infects and enters a host cell, it uses the host cell's enzymes and ribosomes to make thousands of copies of the virus. New viruses eventually leave and then infect other cells. HIV infection of a host cell begins with the binding of HIV glycoproteins to receptors in the host cell's plasma membrane. This causes the cell to transport the virus into its cytoplasm via receptor-mediated endocytosis.

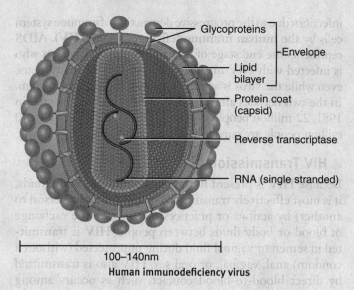

100–140nm
Human immunodeficiency virus

Once inside the host cell, HIV sheds its protein coat, and a viral enzyme called **reverse transcriptase** reads the viral RNA strand and makes a DNA copy. The viral DNA copy then becomes integrated into the host cell's DNA. Thus, the viral DNA is duplicated along with the host cell's DNA during normal cell division. In addition, the viral DNA can cause the infected cell to begin producing millions of copies of viral RNA and to assemble new protein coats for each copy. The new HIV copies bud off from the cell's plasma membrane and circulate in the blood to infect other cells.

HIV mainly damages helper T cells, and it does so in various ways. Over 10 billion viral copies may be produced each day. The viruses bud so rapidly from an infected cell's plasma membrane that cell lysis eventually occurs. In addition, the body's defenses attack the infected cells, killing them but not all the viruses they harbor. In most HIV-infected individuals, helper T cells are initially replaced as fast as they are destroyed. After several years, however, the body's ability to replace helper T cells is slowly exhausted, and the number of helper T cells in circulation progressively declines.

■ Signs, Symptoms, and Diagnosis of HIV Infection

Soon after being infected with HIV, most people experience a brief flu-like illness. Common signs and symptoms are fever, fatigue, rash, headache, joint pain, sore throat, and swollen lymph nodes. About 50% of infected people also experience night sweats. As early as three to four weeks after HIV infection, plasma cells begin secreting antibodies against HIV. These antibodies are detectable in blood plasma and form the basis for some of the screening tests for HIV. When people test "HIV-positive," it usually means they have antibodies to HIV antigens in their bloodstream.

■ Progression to AIDS

After a period of 2 to 10 years, HIV destroys enough helper T cells that most infected people begin to experience symptoms of immunodeficiency. HIV-infected people commonly have enlarged lymph nodes and experience persistent fatigue, involuntary weight loss, night sweats,

skin rashes, diarrhea, and various lesions of the mouth and gums. In addition, the virus may begin to infect neurons in the brain, affecting the person's memory and producing visual disturbances.

As the immune system slowly collapses, an HIV-infected person becomes susceptible to a host of *opportunistic infections*. These are diseases caused by microorganisms that are normally held in check but now proliferate because of the defective immune system. AIDS is diagnosed when the helper T cell count drops below 200 cells per microliter (= cubic millimeter) of blood or when opportunistic infections arise, whichever occurs first. In time, opportunistic infections usually are the cause of death.

■ Treatment of HIV Infection

At present, infection with HIV cannot be cured. Vaccines designed to block new HIV infections and to reduce the viral load (the number of copies of HIV RNA in a microliter of blood plasma) in those who are already infected are in clinical trials. Meanwhile, two categories of drugs have proved successful in extending the life of many HIV-infected people:

1. **Reverse transcriptase inhibitors** interfere with the action of the reverse transcriptase enzyme, the enzyme that the virus uses to convert its RNA into a DNA copy. Among the drugs in this category are zidovudine (ZDV, previously called AZT), didanosine (ddI), and stavudine (d4T®). Trizivir, approved in 2000 for treatment of HIV infection, combines three reverse transcriptase inhibitors in one pill.
2. **Protease inhibitors** interfere with the action of protease, a viral enzyme that cuts proteins into pieces to assemble the protein coat of newly produced HIV particles. Drugs in this category include nelfinavir, saquinavir, ritonavir, and indinavir.

In 1996, physicians treating HIV-infected patients widely adopted *highly active antiretroviral therapy (HAART)*—a combination of two differently acting reverse transcriptase inhibitors and one protease inhibitor. Most HIV-infected individuals receiving HAART experience a drastic reduction in viral load and an increase in the number of helper T cells in their blood. Not only does HAART delay the progression of HIV infection to AIDS, but many individuals with AIDS have seen the remission or disappearance of opportunistic infections and an apparent return to health. Unfortunately, HAART is very costly (exceeding $10,000 per year), the dosing schedule is grueling, and not all people can tolerate the toxic side effects of these drugs. Although HIV may virtually disappear from the blood with drug treatment (and thus a blood test may be "negative" for HIV), the virus typically still lurks in various lymphatic tissues. In such cases, the infected person can still transmit the virus to another person.

■ Allergic Reactions

A person who is overly reactive to a substance that is tolerated by most other people is said to be **allergic** or

hypersensitive. Whenever an allergic reaction takes place, some tissue injury occurs. The antigens that induce an allergic reaction are called **allergens.** Common allergens include certain foods (milk, peanuts, shellfish, eggs), antibiotics (penicillin, tetracycline), vaccines (pertussis, typhoid), venoms (honeybee, wasp, snake), cosmetics, chemicals in plants such as poison ivy, pollens, dust, molds, iodine-containing dyes used in certain x-ray procedures, and even microbes.

There are four basic types of hypersensitivity reactions: type I (anaphylactic), type II (cytotoxic), type III (immune-complex), and type IV (cell-mediated). The first three are antibody-mediated immune responses; the last is a cell-mediated immune response.

Type I (anaphylactic) reactions are the most common and occur within a few minutes after a person sensitized to an allergen is reex-posed to it. In response to the first exposure to certain allergens, some people produce IgE antibodies that bind to the surface of mast cells and basophils. The next time the same allergen enters the body, it attaches to the IgE antibodies already present. In response, the mast cells and basophils release histamine, prostaglandins, leukotrienes, and kinins. Collectively, these mediators cause vasodilation, increased blood capillary permeability, increased smooth muscle contraction in the airways of the lungs, and increased mucus secretion. As a result, a person may experience inflammatory responses, difficulty in breathing through the constricted airways, and a runny nose from excess mucus secretion. In **anaphylactic shock,** which may occur in a susceptible individual who has just received a triggering drug or been stung by a wasp, wheezing and shortness of breath as airways constrict are usually accompanied by shock due to vasodilation and fluid loss from blood. This life-threatening emergency is usually treated by injecting epinephrine to dilate the airways and strengthen the heartbeat.

Type II (cytotoxic) reactions are caused by antibodies (IgG or IgM) directed against antigens on a person's blood cells (red blood cells, lymphocytes, or platelets) or tissue cells. The reaction of antibodies and antigens usually leads to activation of complement. Type II reactions, which may occur in incompatible blood transfusion reactions, damage cells by causing lysis.

Type III (immune-complex) reactions involve antigens, antibodies (IgA or IgM), and complement. When certain ratios of antigen to antibody occur, the immune complexes are small enough to escape phagocytosis, but they become trapped in the basement membrane under the endothelium of blood vessels, where they activate complement and cause inflammation. Glomerulonephritis and rheumatoid arthritis (RA) arise in this way.

Type IV (cell-mediated) reactions or **delayed hypersensitivity reactions** usually appear 12–72 hours after exposure to an allergen. Type IV reactions occur when allergens are taken up by antigen-presenting cells (such as Langerhans cells in the skin) that migrate to lymph nodes and present the allergen to T cells, which then proliferate. Some of the new T cells return to the site of allergen entry into the body, where they produce gamma-interferon, which activates macrophages, and tumor necrosis factor, which stimulates an inflammatory response. Intracellular bacteria such as *Mycobacterium tuberculosis* trigger this type of cell-mediated immune response, as do certain haptens, such as poison ivy toxin. The skin test for tuberculosis also is a delayed hypersensitivity reaction.

Autoimmune Diseases

In an **autoimmune disease** (aw-tō-i-MŪN) or **autoimmunity,** the immune system fails to display self-tolerance and attacks the person's own tissues. Autoimmune diseases usually arise in early adulthood and are common, afflicting an estimated 5% of adults in North America and Europe. Females suffer autoimmune diseases twice as often as males. Self-reactive B cells and T cells normally are deleted or undergo anergy during negative selection. Apparently, this process is not 100% effective. Under the influence of unknown environmental triggers and certain genes that make some people more susceptible, self-tolerance breaks down, leading to activation of self-reactive clones of T cells and B cells. These cells then generate cell-mediated or antibody-mediated immune responses against self-antigens.

A variety of mechanisms produce different autoimmune diseases. Some involve production of **autoantibodies,** antibodies that bind to and stimulate or block self-antigens. For example, autoantibodies that mimic TSH (thyroid-stimulating hormone) are present in Graves disease and stimulate secretion of thyroid hormones (thus producing hyperthyroidism); autoantibodies that bind to and block acetylcholine receptors cause the muscle weakness characteristic of myasthenia gravis. Other autoimmune diseases involve activation of cytotoxic T cells that destroy certain body cells. Examples include type 1 diabetes mellitus, in which T cells attack the insulin-producing pancreatic beta cells, and multiple sclerosis (MS), in which T cells attack myelin sheaths around axons of neurons. Inappropriate activation of helper T cells or excessive production of gamma-interferon also occur in certain autoimmune diseases. Other autoimmune disorders include rheumatoid arthritis (RA), systemic lupus erythematosus (SLE), rheumatic fever, hemolytic and pernicious anemias, Addison's disease, Hashimoto's thyroiditis, and ulcerative colitis.

Therapies for various autoimmune diseases include removal of the thymus gland (thymectomy), injections of beta interferon, immunosuppressive drugs, and plasmapheresis, in which the person's blood plasma is filtered to remove antibodies and antigen–antibody complexes.

Cancer Immunology

Although the immune system responds to cancerous cells, often immunity provides inadequate protection, as evidenced by the number of people dying each year from cancer. Over the past 25 years, considerable research has focused on *cancer immunology,* the study of ways to use immune responses for detecting, monitoring, and treating cancer. For example, some tumors of the colon release *carcinoembryonic antigen*

(*CEA*) into the blood, and prostate cancer cells release *prostate-specific antigen* (*PSA*). Detecting these antigens in blood does not provide definitive diagnosis of cancer, because both antigens are also released in certain noncancerous conditions. However, high levels of cancer-related antigens in the blood often do indicate the presence of a malignant tumor.

Finding ways to induce our immune system to mount vigorous attacks against cancerous cells has been an elusive goal. Many different techniques have been tried, with only modest success. In one method, inactive lymphocytes are removed in a blood sample and cultured with interleukin-2. The resulting *lymphokine-activated killer* (*LAK*) *cells* are then transfused back into the patient's blood. Although LAK cells have produced dramatic improvement in a few cases, severe complications affect most patients. In another method, lymphocytes procured from a small biopsy sample of a tumor are cultured with interleukin-2. After their proliferation in culture, *such tumor-infiltrating lymphocytes* (*TILs*) are reinjected. About a quarter of patients with malignant melanoma and renal-cell carcinoma who received TIL therapy showed significant improvement. The many studies currently underway provide reason to hope that immune-based methods will eventually lead to cures for cancer.

Chronic fatigue syndrome

Chronic fatigue syndrome (*CFS*) is a disorder, usually occurring in young adults and primarily in females, characterized by (1) extreme fatigue that impairs normal activities for at least 6 months and (2) the absence of other known diseases (cancer, infections, drug abuse, toxicity, or psychiatric disorders) that might produce similar symptoms.

Infectious Mononucleosis

Infectious mononucleosis or "mono" is a contagious disease caused by the *Epstein–Barr virus* (*EBV*). It occurs mainly in children and young adults, and more often in females than in males. The virus most commonly enters the body through intimate oral contact such as kissing, which accounts for its common name, the "kissing disease." EBV then multiplies in lymphatic tissues and filters into the blood, where it infects and multiplies in B cells, the primary host cells. Because of this infection, the B cells become so enlarged and abnormal in appearance that they resemble monocytes, the primary reason for the term **mononucleosis.** In addition to an elevated white blood cell count with an abnormally high percentage of lymphocytes, signs and symptoms include fatigue, headache, dizziness, sore throat, enlarged and tender lymph nodes, and fever. There is no cure for infectious mononucleosis, but the disease usually runs its course in a few weeks.

Lymphomas

Lymphomas (lim-FŌ-mas; *lymph-* = clear water; *-oma* = tumor) are cancers of the lymphatic organs, especially the lymph nodes. Most have no known cause. The two main types of lymphomas are Hodgkin disease and non-Hodgkin lymphoma.

Hodgkin disease (HD) is characterized by a painless, nontender enlargement of one or more lymph nodes, most commonly in the neck, chest, and axilla. If the disease has metastasized from these sites, fever, night sweats, weight loss, and bone pain also occur. HD primarily affects individuals between ages 15 and 35 and those over 60, and it is more common in males. If diagnosed early, HD has a 90–95% cure rate.

Non-Hodgkin lymphoma (NHL), which is more common than HD, occurs in all age groups, the incidence increasing with age to a maximum between ages 45 and 70. NHL may start the same way as HD but may also include an enlarged spleen, anemia, and general malaise. Up to half of all individuals with NHL are cured or survive for a lengthy period. Treatment options for both HD and NHL include radiation therapy, chemotherapy, and bone marrow transplantation.

Severe combined immunodeficiency disease

Severe combined immunodeficiency disease (SCID) is a rare inherited disorder in which both B cells and T cells are missing or inactive. Scientists have now identified mutations in several genes that are responsible for some types of SCID. In some cases, an infusion of red bone marrow cells from a sibling having very similar MHC (HLA) antigens can provide normal stem cells that give rise to normal B and T cells. The result can be a complete cure. Less than 30% of afflicted patients, however, have a compatible sibling who could serve as a donor.

Systemic Lupus Erythematosus

Systemic lupus erythematosus (er-e′-thēm-a-TŌ-sus), **SLE,** or simply **lupus** (*lupus* = wolf) is a chronic autoimmune, inflammatory disease that affects multiple body systems. Lupus is characterized by periods of active disease and remission; symptoms range from mild to life-threatening. Lupus most often develops between ages 15 and 44 and is 10–15 times more common in females than males. It is also 2–3 times more common in African-Americans, Hispanics, Asian-Americans, and Native Americans than in European-Americans. Although the cause of SLE is not known, both a genetic predisposition to the disease and environmental factors (infections, antibiotics, ultraviolet light, stress, and hormones) may trigger it. Sex hormones appear to influence the development of SLE. The disorder often occurs in females who exhibit extremely low levels of androgens.

Signs and symptoms of SLE include joint pain, muscle pain, chest pain with deep breaths, headaches, pale or purple fingers or toes, kidney dysfunction, low blood cell count, nerve or brain dysfunction, slight fever, fatigue, oral ulcers, weight loss, swelling in the legs or around the eyes, enlarged lymph nodes and spleen, photosensitivity, rapid loss of large amounts of scalp hair, and sometimes an eruption across the bridge of the nose and cheeks called a "butterfly rash." The erosive nature of some of the SLE skin lesions was thought to resemble the damage inflicted by the bite of a wolf—thus, the term *lupus*.

Two immunological features of SLE are excessive activation of B cells and inappropriate production of autoantibodies against DNA (anti-DNA antibodies) and other components of cellular nuclei such as histone proteins. Triggers of B cell activation are thought to include various chemicals and drugs, viral and bacterial antigens, and exposure to sunlight. Circulating complexes of abnormal autoantibodies and their "antigens" cause damage in tissues throughout the body. Kidney damage occurs as the complexes become trapped in the basement membrane of kidney capillaries, obstructing blood filtering. Renal failure is the most common cause of death. There is no cure for lupus, but drug therapy can minimize symptoms, reduce inflammation, and forestall flare-ups. The most commonly used lupus medications are pain relievers (nonsteroidal anti-inflammatory drugs such as aspirin and ibuprofen), antimalarial drugs (hydroxychloroquine), and corticosteroids (prednisone and hydrocortisone).

Additional Clinical Considerations

■ Abscesses and Ulcers

If pus cannot drain out of an inflamed region, the result is an **abscess**—an excessive accumulation of pus in a confined space. Common examples are pimples and boils. When superficial inflamed tissue sloughs off the surface of an organ or tissue, the resulting open sore is called an **ulcer**. People with poor circulation—for instance, diabetics with advanced atherosclerosis—are susceptible to ulcers in the tissues of their legs. These ulcers, which are called stasis ulcers, develop because of poor oxygen and nutrient supply to tissues that then become very susceptible to even a very mild injury or an infection.

■ Metastasis through Lymphatic Vessels

Metastasis (me-TAS-ta-sis; *meta-* = beyond; *stasis* = to stand), the spread of a disease from one part of the body to another, can occur via lymphatic vessels. All malignant tumors eventually metastasize. Cancer cells may travel in the blood or lymph and establish new tumors where they lodge. When metastasis occurs via lymphatic vessels, secondary tumor sites can be predicted according to the direction of lymph flow from the primary tumor site. Cancerous lymph nodes feel enlarged, firm, nontender, and fixed to underlying structures. By contrast, most lymph nodes that are enlarged due to an infection are softer, tender, and movable.

■ Microbial Evasion of Phagocytosis

Some microbes, such as the bacteria that cause pneumonia, have extracellular structures called capsules that prevent adherence. This makes it physically difficult for phagocytes to engulf the microbes. Other microbes, such as the toxin-producing bacteria that cause one kind of food poisoning, may be ingested but not killed; instead, the toxins they produce (leukocidins) may kill the phagocytes by causing the release of the phagocyte's own lysosomal enzymes into its cytoplasm. Still other microbes—such as the bacteria that cause tuberculosis—inhibit fusion of phagosomes and lysosomes and thus prevent exposure of the microbes to lysosomal enzymes. These bacteria apparently can also use chemicals in their cell walls to counter the effects of lethal oxidants produced by phagocytes. Subsequent multiplication of the microbes within phagosomes may eventually destroy the phagocyte.

■ Monoclonal Antibodies

The antibodies produced against a given antigen by plasma cells can be harvested from an individual's blood. However, because an antigen typically has many epitopes, several different clones of plasma cells produce different antibodies against the antigen. If a single plasma cell could be isolated and induced to proliferate into a clone of identical plasma cells, then a large quantity of identical antibodies could be produced. Unfortunately, lymphocytes and plasma cells are difficult to grow in culture, so scientists sidestepped this difficulty by fusing B cells with tumor cells that grow easily and proliferate endlessly. The resulting hybrid cell is called a **hybridoma** (hī-bri-DŌ-ma). Hybridomas are long-term sources of large quantities of pure, identical antibodies, called **monoclonal antibodies (MAbs)** because they come from a single clone of identical cells. One clinical use of monoclonal antibodies is for measuring levels of a drug in a patient's blood. Other uses include the diagnosis of strep throat, pregnancy, allergies, and diseases such as hepatitis, rabies, and some sexually transmitted diseases. MAbs have also been used to detect cancer at an early stage and to ascertain the extent of metastasis. They may also be useful in preparing vaccines to counteract the rejection associated with transplants, to treat autoimmune diseases, and perhaps to treat AIDS.

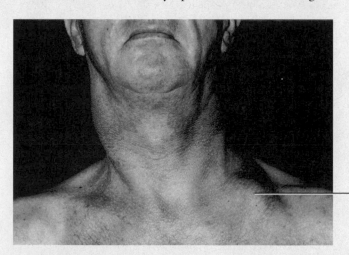

Cancerous lymph node

Cancerous lymph node in neck due to metastasis from a carrier in the pharynx (throat)

■ Ruptured Spleen

The spleen is the organ most often damaged in cases of abdominal trauma. Severe blows over the inferior left chest or superior abdomen can fracture the protecting ribs. Such crushing injury may result in a **ruptured spleen,** which causes significant hemorrhage and shock. Prompt removal of the spleen, called a **splenectomy,** is needed to prevent death due to bleeding. Other structures, particularly red bone marrow and the liver, can take over some functions normally carried out by the spleen. Immune functions, however, decrease in the absence of a spleen. The spleen's absence also places the patient at higher risk for **sepsis** (a blood infection) due to loss of the filtering and phagocytic functions of the spleen. To reduce the risk of sepsis, patients who have undergone a splenectomy take prophylactic (preventive) antibiotics before any invasive procedures. ■

Medical Terminology

Adenitis (ad′-e-NĪ-tis; *aden-* = gland; *-itis* = inflammation of) Enlarged, tender, and inflamed lymph nodes resulting from an infection.

Allograft (AL-ō-graft; *allo-* = other) A transplant between genetically distinct individuals of the same species. Skin transplants from other people and blood transfusions are allografts.

Autograft (AW-tō-graft; *auto-* = self) A transplant in which one's own tissue is grafted to another part of the body (such as skin grafts for burn treatment or plastic surgery).

Gamma globulin (GLOB-ū-lin) Suspension of immunoglobulins from blood consisting of antibodies that react with a specific pathogen. It is prepared by injecting the pathogen into animals, removing blood from the animals after antibodies have been produced, isolating the antibodies, and injecting them into a human to provide short-term immunity.

Hypersplenism (hī-per-SPLĒN-izm; *hyper-* = over) Abnormal splenic activity due to splenic enlargement and associated with an increased rate of destruction of normal blood cells.

Lymphadenopathy (lim-fad′-e-NOP-a-thē; *lymph-* = clear fluid; *-pathy* = disease) Enlarged, sometimes tender lymph glands as a response to infection; also called *swollen glands.*

Lymphangitis (lim-fan-JĪ-tis; *-itis* = inflammation of) Inflammation of lymphatic vessels.

Lymphedema (lim′-fe-DĒ-ma; *edema* = swelling) Accumulation of lymph in lymphatic vessels, causing painless swelling of a limb.

Splenomegaly (splē′-nō-MEG-a-lē; *mega-* = large) Enlarged spleen.

Tonsillectomy (ton′-si-LEK-tō-mē; *-ectomy* = excision) Removal of a tonsil.

Xenograft (ZEN-ō-graft; *xeno-* = strange or foreign) A transplant between animals of different species. Xenografts from porcine (pig) or bovine (cow) tissue may be used in humans as a physiological dressing for severe burns. Other xenografts include pig heart valves and baboon hearts.

Chapter 11

The Respiratory System

The **respiratory system** provides for gas exchange–intake of oxygen and elimination of carbon dioxide–and the cardiovascular system transports blood containing the gases between the lungs and body cells. Failure of either system disrupts homeostasis by causing rapid death of cells from oxygen starvation and buildup of waste products. In addition, the respiratory system participates in regulating blood pH, contains receptors for the sense of smell, filters inspired air, produces sounds, and rids the body of some water and heat in exhaled air.

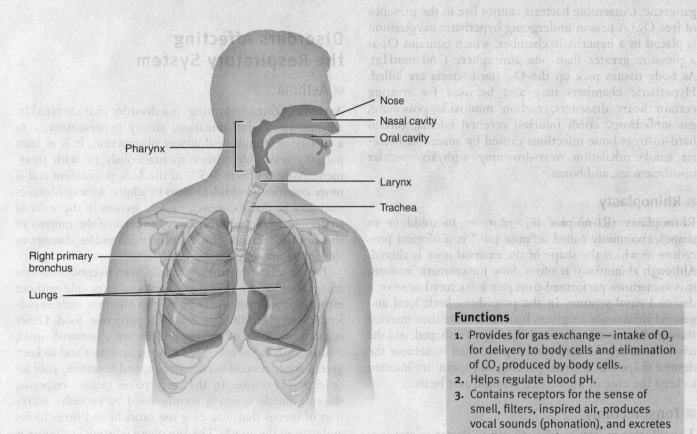

Anterior view showing organs of respiration

- Nose
- Nasal cavity
- Oral cavity
- Pharynx
- Larynx
- Trachea
- Right primary bronchus
- Lungs

Functions

1. Provides for gas exchange—intake of O_2 for delivery to body cells and elimination of CO_2 produced by body cells.
2. Helps regulate blood pH.
3. Contains receptors for the sense of smell, filters, inspired air, produces vocal sounds (phonation), and excretes small amounts of water and heat.

Tests, Procedures, and Techniques

Abdominal thrust maneuver

Abdominal thrust maneuver is a first-aid procedure designed to clear the airways of obstructing objects. It is performed by applying a quick upward thrust between the navel and costal margin that causes sudden elevation of the diaphragm and forceful, rapid expulsion of air in the lungs; this action forces air out the trachea to eject the obstructing object. The abdominal thrust maneuver is also used to expel water from the lungs of near-drowning victims before resuscitation is begun. Previously this was called the **Heimlich maneuver** (HĪM-lik ma-NOO-ver).

Bronchography

Bronchography (bron-KOG-ra-fē) is an imaging technique used to visualize the bronchial tree using x-rays. After an opaque contrast medium is inhaled through an intratracheal catheter, radiographs of the chest in various positions are taken, and the developed film, a **bronchogram** (BRON-kō-gram), provides a picture of the bronchial tree.

Bronchoscopy

Bronchoscopy (bron-KOS-kō-pē) is the visual examination of the bronchi through a **bronchoscope,** an illuminated, flexible tubular instrument that is passed through the mouth (or nose), larynx, and trachea into the bronchi. The examiner can view the interior of the trachea and bronchi to biopsy a tumor, clear an obstructing object or secretions from an airway, take cultures or smears for microscopic examination, stop bleeding, or deliver drugs.

Hyperbaric Oxygenation

A major clinical application of Henry's law is **hyperbaric oxygenation** (*hyper* = over; *baros* = pressure), the use of pressure to cause more O_2 to dissolve in the blood. It is an effective technique in treating patients infected by anaerobic bacteria, such as those that cause tetanus and gangrene. (Anaerobic bacteria cannot live in the presence of free O_2.) A person undergoing hyperbaric oxygenation is placed in a hyperbaric chamber, which contains O_2 at a pressure greater than one atmosphere (760 mmHg). As body tissues pick up the O_2, the bacteria are killed. Hyperbaric chambers may also be used for treating certain heart disorders, carbon monoxide poisoning, gas embolisms, crush injuries, cerebral edema, certain hard-to-treat bone infections caused by anaerobic bacteria, smoke inhalation, near-drowning, asphyxia, vascular insufficiencies, and burns.

Rhinoplasty

Rhinoplasty (RĪ-nō-plas′-tē; *-plasty* = to mold or to shape), commonly called a "nose job," is a surgical procedure in which the shape of the external nose is altered. Although rhinoplasty is often done for cosmetic reasons, it is sometimes performed to repair a fractured nose or a deviated nasal septum. In the procedure, both local and general anesthetics are given. Instruments are then inserted through the nostrils, the nasal cartilage is reshaped, and the nasal bones are fractured and repositioned to achieve the desired shape. An internal packing and splint are inserted to keep the nose in the desired position as it heals.

Tonsillectomy

Tonsillectomy (ton-si-LEK-tō-mē-; *-ektome* = excision) is surgical removal of the tonsils. The procedure is usually performed under general anesthesia on an outpatient basis. Tonsillectomies are performed in individuals who have frequent *tonsillitis* (ton′-si-LĪ-tis), that is, inflammation of the tonsils; tonsils that develop an abcess or tumor; or when the tonsils obstruct breathing during sleep.

Tracheotomy and Intubation

Several conditions may block airflow by obstructing the trachea. For example, the rings of cartilage that support the trachea may collapse due to a crushing injury to the chest, inflammation of the mucous membrane may cause it to swell so much that the airway closes, vomit or a foreign object may be aspirated into it, or a cancerous tumor may protrude into the airway. Two methods are used to reestablish airflow past a tracheal obstruction. If the obstruction is superior to the level of the larynx, a tracheotomy (trā-kē-O-tō-mē; *-tome* = cutting), an operation to make an opening into the trachea, may be performed. In this procedure, also called a *tracheostomy*, a skin incision is followed by a short longitudinal incision into the trachea inferior to the cricoid cartilage. The patient can then breathe through a metal or plastic tracheal tube inserted through the incision. The second method is **intubation,** in which a tube is inserted into the mouth or nose and passed inferiorly through the larynx and trachea. The firm wall of the tube pushes aside any flexible obstruction, and the lumen of the tube provides a passageway for air; any mucus clogging the trachea can be suctioned out through the tube. ∎

Disorders Affecting the Respiratory System

Asthma

Asthma (AZ-ma = panting) is a disorder characterized by chronic airway inflammation, airway hypersensitivity to a variety of stimuli, and airway obstruction. It is at least partially reversible, either spontaneously or with treatment. Asthma affects 3–5% of the U.S. population and is more common in children than in adults. Airway obstruction may be due to smooth muscle spasms in the walls of smaller bronchi and bronchioles, edema of the mucosa of the airways, increased mucus secretion, and/or damage to the epithelium of the airway.

Individuals with asthma typically react to concentrations of agents too low to cause symptoms in people without asthma. Sometimes the trigger is an allergen such as pollen, house dust mites, molds, or a particular food. Other common triggers of asthma attacks are emotional upset, aspirin, sulfiting agents (used in wine and beer and to keep greens fresh in salad bars), exercise, and breathing cold air or cigarette smoke. In the early phase (acute) response, smooth muscle spasm is accompanied by excessive secretion of mucus that may clog the bronchi and bronchioles and worsen the attack. The late phase (chronic) response is characterized by inflammation, fibrosis, edema, and necrosis (death) of bronchial epithelial cells. A host of mediator chemicals, including leukotrienes, prostaglandins, thromboxane, platelet-activating factor, and histamine, take part.

Symptoms include difficult breathing, coughing, wheezing, chest tightness, tachycardia, fatigue, moist skin, and anxiety. An acute attack is treated by giving an inhaled beta$_2$-adrenergic agonist (albuterol) to help relax smooth muscle in the bronchioles and open up the airways. However, long-term therapy of asthma strives to suppress the underlying inflammation. The anti-inflammatory drugs that are used most often are inhaled corticosteroids (glucocorticoids), cromolyn sodium (Intal®), and leukotriene blockers (Accolate®).

Avian influenza

Avian influenza is a respiratory disorder that has resulted in the deaths of hundreds of millions of birds worldwide.

It is usually transmitted from one bird to another bird through their droppings, saliva, and nasal secretions. Currently, avian influenza is difficult to transmit from birds to humans; the few humans who have died from avian influenza have had close contact with infected birds. Avian influenza is also called **bird flu.**

■ Coryza and Influenza

Hundreds of viruses can cause **coryza** (ko-RĪ-za) or the **common cold,** but a group of viruses called *rhinoviruses* is responsible for about 40% of all colds in adults. Typical symptoms include sneezing, excessive nasal secretion, dry cough, and congestion. The uncomplicated common cold is not usually accompanied by a fever. Complications include sinusitis, asthma, bronchitis, ear infections, and laryngitis. Recent investigations suggest an association between emotional stress and the common cold. The higher the stress level, the greater the frequency and duration of colds.

Influenza (flu) is also caused by a virus. Its symptoms include chills, fever (usually higher than 101°F = 39°C), headache, and muscular aches. Influenza can become life-threatening and may develop into pneumonia. It is important to recognize that influenza is a respiratory disease, not a gastrointestinal (GI) disease. Many people mistakenly report having "the flu" when they are suffering from a GI illness.

Sneezing is one symptom of the common cold

■ Chronic Bronchitis

Chronic bronchitis is a disorder characterized by excessive secretion of bronchial mucus accompanied by a productive cough (sputum is raised) that lasts for at least three months of the year for two successive years. Cigarette smoking is the leading cause of chronic bronchitis. Inhaled irritants lead to chronic inflammation with an increase in the size and number of mucous glands and goblet cells in the airway epithelium. The thickened and excessive mucus produced narrows the airway and impairs ciliary function. Thus, inhaled pathogens become embedded in airway secretions and multiply rapidly. Besides a productive cough, symptoms of chronic bronchitis are shortness of breath, wheezing, cyanosis, and pulmonary hypertension. Treatment for chronic bronchitis is similar to that for emphysema.

■ Chronic Obstructive Pulmonary Disease

Chronic obstructive pulmonary disease (COPD) is a type of respiratory disorder characterized by chronic and recurrent obstruction of airflow, which increases airway resistance. COPD affects about 30 million Americans and is the fourth leading cause of death behind heart disease, cancer, and cerebrovascular disease. The principal types of COPD are emphysema and chronic bronchitis. In most cases, COPD is preventable because its most common cause is cigarette smoking or breathing secondhand smoke. Other causes include air pollution, pulmonary infection, occupational exposure to dusts and gases, and genetic factors. Because men, on average, have more years of exposure to cigarette smoke than women, they are twice as likely as women to suffer from COPD; still, the incidence of COPD in women has risen sixfold in the past 50 years, a reflection of increased smoking among women.

■ Cystic Fibrosis

Cystic fibrosis (CF) is an inherited disease of secretory epithelia that affects the airways, liver, pancreas, small intestine, and sweat glands. It is the most common lethal genetic disease in whites: 5% of the population are thought to be genetic carriers. The cause of cystic fibrosis is a genetic mutation affecting a transporter protein that carries chloride ions across the plasma membranes of many epithelial cells. Because dysfunction of sweat glands causes perspiration to contain excessive sodium chloride (salt), measurement of the excess chloride is one index for diagnosing CF. The mutation also disrupts the normal functioning of several organs by causing ducts within them to become obstructed by thick mucus secretions that do not drain easily from the passageways. Buildup of these secretions leads to inflammation and replacement of injured cells with connective tissue that further blocks the ducts. Clogging and infection of the airways leads to difficulty in breathing and eventual destruction of lung tissue. Lung disease accounts for most deaths from CF. Obstruction of small bile ducts in the liver interferes with digestion and disrupts liver function; clogging of pancreatic ducts prevents digestive enzymes from reaching the small intestine. Because pancreatic juice contains the main fat-digesting enzyme, the person fails to absorb fats or fat-soluble vitamins and thus suffers from vitamin A, D, and K deficiency diseases. With respect to the reproductive systems, blockage of the ductus (vas) deferens leads to infertility in males; the formation of dense mucus plugs in the vagina restricts the entry of sperm into the uterus and can lead to infertility in females.

A child suffering from cystic fibrosis is given pancreatic extract and large doses of vitamins A, D, and K. The recommended diet is high in calories, fats, and proteins, with vitamin supplementation and liberal use of salt. One of the newest treatments for CF is heart–lung transplants.

■ Emphysema

Emphysema (em-fi-SĒ-ma = blown up or full of air) is a disorder characterized by destruction of the walls of the alveoli, producing abnormally large air spaces that remain filled with air during exhalation. With less surface area for gas exchange, O_2 diffusion across the damaged respiratory membrane is reduced. Blood O_2 level is somewhat lowered, and any mild exercise that raises the O_2 requirements of the cells leaves the patient breathless. As increasing numbers of alveolar walls are damaged, lung elastic recoil decreases due to loss of elastic fibers, and an increasing amount of air becomes trapped in the lungs at the end of exhalation. Over several years, added exertion during inhalation increases the size of the chest cage, resulting in a "barrel chest."

Emphysema is generally caused by a long-term irritation; cigarette smoke, air pollution, and occupational exposure to industrial dust are the most common irritants. Some destruction of alveolar sacs may be caused by an enzyme imbalance. Treatment consists of cessation of smoking, removal of other environmental irritants, exercise training under careful medical supervision, breathing exercises, use of bronchodilators, and oxygen therapy.

■ Hypoxia

Hypoxia (hī-POK-sē-a; *hypo-* = under) is a deficiency of O_2 at the tissue level. Based on the cause, we can classify hypoxia into four types, as follows:

1. **Hypoxic hypoxia** is caused by a low P_{O_2} in arterial blood as a result of high altitude, airway obstruction, or fluid in the lungs.
2. In **anemic hypoxia,** too little functioning hemoglobin is present in the blood, which reduces O_2 transport to tissue cells. Among the causes are hemorrhage, anemia, and failure of hemoglobin to carry its normal complement of O_2, as in carbon monoxide poisoning.
3. In **ischemic hypoxia,** blood flow to a tissue is so reduced that too little O_2 is delivered to it, even though P_{O_2} and oxyhemoglobin levels are normal.
4. In **histotoxic hypoxia,** the blood delivers adequate O_2 to tissues, but the tissues are unable to use it properly because of the action of some toxic agent. One cause is cyanide poisoning, in which cyanide blocks an enzyme required for the use of O_2 during ATP synthesis.

■ Laryngitis and Cancer of the larynx

Laryngitis is an inflammation of the larynx that is most often caused by a respiratory infection or irritants such as cigarette smoke. Inflammation of the vocal folds causes hoarseness or loss of voice by interfering with the contraction of the folds or by causing them to swell to the point where they cannot vibrate freely. Many long-term smokers acquire a permanent hoarseness from the damage done by chronic inflammation. **Cancer of the larynx** is found almost exclusively in individuals who smoke. The condition is characterized by hoarseness, pain on swallowing, or pain radiating to an ear. Treatment consists of radiation therapy and/or surgery.

■ Lung Cancer

In the United States, **lung cancer** is the leading cause of cancer death in both males and females, accounting for 160,000 deaths annually. At the time of diagnosis, lung cancer is usually well advanced, with distant metastases present in about 55% of patients, and regional lymph node involvement in an additional 25%. Most people with lung cancer die within a year of the initial diagnosis; the overall survival rate is only 10–15%. Cigarette smoke is the most common cause of lung cancer. Roughly 85% of lung cancer cases are related to smoking, and the disease is 10 to 30 times more common in smokers than nonsmokers. Exposure to secondhand smoke is also associated with lung cancer and heart disease. In the United States, secondhand smoke causes an estimated 4000 deaths a year from lung cancer, and nearly 40,000 deaths a year from heart disease. Other causes of lung cancer are ionizing radiation and inhaled irritants, such as asbestos and radon gas. Emphysema is a common precursor to the development of lung cancer.

The most common type of lung cancer, **bronchogenic carcinoma,** starts in the epithelium of the bronchial tubes. Bronchogenic tumors are named based on where they arise. For example, *adenocarcinomas* develop in peripheral areas of the lungs from bronchial glands and alveolar cells, *squamous cell carcinomas* develop from the epithelium of larger bronchial tubes, and *small (oat) cell carcinomas* develop from epithelial cells in primary bronchi near the hilum of the lungs and tend to involve the mediastinum early on. Depending on the type of bronchogenic tumors, they may be aggressive, locally invasive, and undergo widespread metastasis. The tumors begin as epithelial lesions that grow to form masses that obstruct the bronchial tubes or invade adjacent lung tissue. Bronchogenic carcinomas metastasize to lymph nodes, the brain, bones, liver, and other organs.

Symptoms of lung cancer are related to the location of the tumor. These may include a chronic cough, spitting blood from the respiratory tract, wheezing, shortness of breath, chest pain, hoarseness, difficulty swallowing, weight loss, anorexia, fatigue, bone pain, confusion, problems with balance, headache, anemia, thrombocytopenia, and jaundice. For some people, there are relatively few or no dramatic symptoms.

Treatment consists of partial or complete surgical removal of a diseased lung (pulmonectomy), radiation therapy, and chemotherapy.]

Cigarette smoke is the most common cause of lung cancer

Pneumonia

Pneumonia is an acute infection or inflammation of the alveoli. It is the most common infectious cause of death in the United States, where an estimated 4 million cases occur annually. When certain microbes enter the lungs of susceptible individuals, they release damaging toxins, stimulating inflammation and immune responses that have damaging side effects. The toxins and immune response damage alveoli and bronchial mucous membranes; inflammation and edema cause the alveoli to fill with fluid, interfering with ventilation and gas exchange.

The most common cause of pneumonia is the pneumococcal bacterium *Streptococcus pneumoniae*, but other microbes may also cause pneumonia. Those who are most susceptible to pneumonia are the elderly, infants, immunocompromised individuals (AIDS or cancer patients, or those taking immunosuppressive drugs), cigarette smokers, and individuals with an obstructive lung disease. Most cases of pneumonia are preceded by an upper respiratory infection that often is viral. Individuals then develop fever, chills, productive or dry cough, malaise, chest pain, and sometimes dyspnea (difficult breathing) and hemoptysis (spitting blood).

Treatment may involve antibiotics, bronchodilators, oxygen therapy, increased fluid intake, and chest physiotherapy (percussion, vibration, and postural drainage).

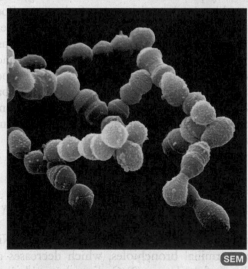

Streptococcus pneumoniae, the most common cause of pneumonia

Respiratory Distress Syndrome

Respiratory distress syndrome (RDS) is a breathing disorder of premature newborns in which the alveoli do not remain open due to a lack of surfactant. Recall that surfactant reduces surface tension and is necessary to prevent the collapse of alveoli during exhalation. The more premature the newborn, the greater the chance that RDS will develop. The condition is also more common in infants whose mothers have diabetes, in males, and occurs more often in European Americans than African Americans. Symptoms of RDS include labored and irregular breathing, flaring of the nostrils during inhalation, grunting during exhalation, and perhaps a blue skin color. Besides the symptoms, RDS is diagnosed on the basis of chest radiographs and a blood test. A newborn with mild RDS may require only supplemental oxygen administered through an oxygen hood or through a tube placed in the nose. In severe cases oxygen may be delivered by continuous positive airway pressure (CPAP) through tubes in the nostrils or a mask on the face. In such cases surfactant may be administered directly into the lungs.

Severe Acute Respiratory Syndrome

Severe acute respiratory syndrome (SARS) is an example of an *emerging infectious disease*, that is, a disease that is new or changing. Other examples of emerging infectious diseases are West Nile encephalitis, mad cow disease, and AIDS. SARS first appeared in Southern China in late 2002 and has subsequently spread worldwide. It is a respiratory illness caused by a new variety of coronavirus. Symptoms of SARS include fever, malaise, muscle aches, nonproductive (dry) cough, difficulty in breathing, chills, headache, and diarrhea. About 10–20% of patients require mechanical ventilation and in some cases death may result. The disease is primarily spread through person-to-person contact. There is no effective treatment for SARS and the death rate is 5–10%, usually among the elderly and in persons with other medical problems.

Sudden Infant Death Syndrome

Sudden infant death syndrome (SIDS) is the sudden, unexpected death of an apparently healthy infant during sleep. It rarely occurs before 2 weeks or after 6 months of age, with the peak incidence between the second and fourth months. SIDS is more common in premature infants, male babies, low-birth-weight babies, babies of drug users or smokers, babies who have stopped breathing and have had to be resuscitated, babies with upper respiratory tract infections, and babies who have had a sibling die of SIDS. African American and Native American babies are at higher risk. The exact cause of SIDS is unknown. However, it may be due to an abnormality in the mechanisms that control respiration or low levels of oxygen in the blood. SIDS may also be linked to hypoxia while sleeping in a prone position (on the stomach) and the rebreathing of exhaled air trapped in a depression of a mattress. It is recommended that for the first six months infants be placed on their backs for sleeping ("back to sleep").

Tuberculosis

The bacterium *Mycobacterium tuberculosis* produces an infectious, communicable disease called **tuberculosis (TB)** that most often affects the lungs and the pleurae but may involve other parts of the body. Once the bacteria are inside the lungs, they multiply and cause inflammation, which stimulates neutrophils and macrophages to migrate to the area and engulf the bacteria to prevent their spread. If the immune system is not impaired, the bacteria remain dormant for life, but impaired immunity may enable the bacteria to escape into blood and lymph to infect other organs. In many people, symptoms—fatigue, weight loss, lethargy, anorexia, a low-grade fever, night sweats, cough, dyspnea, chest pain, and hemoptysis—do not develop until the disease is advanced.

During the past several years, the incidence of TB in the United States has risen dramatically. Perhaps the single most important factor related to this increase is the spread of the human immunodeficiency virus (HIV). People infected with HIV are much more likely to develop tuberculosis because their immune systems are impaired. Among the other factors that have contributed to the increased number of cases are homelessness, increased drug abuse, increased immigration from countries with a high prevalence of tuberculosis, increased crowding in housing among the poor, and airborne transmission of tuberculosis in prisons and shelters. In addition, recent outbreaks of tuberculosis involving multi-drug-resistant strains of *Mycobacterium tuberculosis* have occurred because patients fail to complete their antibiotic and other treatment regimens. TB is treated with the medication isoniazid. ■

Additional Clinical Considerations

■ Asbestos-related Diseases

Asbestos-related diseases are serious lung disorders that develop as a result of inhaling asbestos particles decades earlier. When asbestos particles are inhaled, they penetrate lung tissue. In response, white blood cells attempt to destroy them by phagocytosis. However, the fibers usually destroy the white blood cells and scarring of lung tissue may follow. Asbestos-related diseases include **asbestosis** (widespread scarring of lung tissue), **diffuse pleural thickening** (thickening of the pleurae), and **mesothelioma** (cancer of the pleurae or, less commonly, the peritoneum).

■ Carbon Monoxide Poisoning

Carbon monoxide (CO) is a colorless and odorless gas found in exhaust fumes from automobiles, gas furnaces, and space heaters and in tobacco smoke. It is a byproduct of the combustion of carbon-containing materials such as coal, gas, and wood. CO binds to the heme group of hemoglobin, just as O_2 does, except that the binding of carbon monoxide to hemoglobin is over 200 times as strong as the binding of O_2 to hemoglobin. Thus, at a concentration as small as 0.1% (P_{CO} = 0.5 mmHg), CO will combine with half the available hemoglobin molecules and reduce the oxygen-carrying capacity of the blood by 50 %. Elevated blood levels of CO cause **carbon monoxide poisoning,** which can cause the lips and oral mucosa to appear bright, cherry-red (the color of hemoglobin with carbon monoxide bound to it). Without prompt treatment, carbon monoxide poisoning is fatal. It is possible to rescue a victim of CO poisoning by administering pure oxygen, which speeds up the separation of carbon monoxide from hemoglobin.

■ Pneumothorax and Hemothorax

In certain conditions, the pleural cavities may fill with air (**pneumothorax;** *pneumo-* = air or breath), blood (**hemothorax**), or pus. Air in the pleural cavities, most commonly introduced in a surgical opening of the chest or as a result of a stab or gunshot wound, may cause the lungs to collapse. This collapse of a part of a lung, or rarely an entire lung, is called **atelectasis** (at′-e-LEK-ta-sis; *ateles-* = incomplete; *-ectasis-* = expansion). The goal of treatment is the evacuation of air (or blood) from the pleural space, which allows the lung to reinflate. A small pneumothorax may resolve on its own, but it is often necessary to insert a chest tube to assist in evacuation.

■ Pulmonary Edema

Pulmonary edema is an abnormal accumulation of fluid in the interstitial spaces and alveoli of the lungs. The edema may arise from increased permeability of the pulmonary capillaries (pulmonary origin) or increased pressure in the pulmonary capillaries (cardiac origin); the latter cause may coincide with congestive heart failure. The most common symptom is dyspnea. Others include wheezing, tachypnea (rapid breathing rate), restlessness, a feeling of suffocation, cyanosis, pallor (paleness), diaphoresis (excessive perspiration), and pulmonary hypertension. Treatment consists of administering oxygen, drugs that dilate the bronchioles and lower blood pressure, diuretics to rid the body of excess fluid, and drugs that correct acid–base imbalance; suctioning of airways; and mechanical ventilation. One of the recent culprits in the development of pulmonary edema was found to be "phen-fen" diet pills.

■ The Effect of Smoking on Respiratory Efficiency

Smoking may cause a person to become easily "winded" during even moderate exercise because several factors decrease respiratory efficiency in smokers: (1) Nicotine constricts terminal bronchioles, which decreases airflow into and out of the lungs. (2) Carbon monoxide in smoke binds to hemoglobin and reduces its oxygen-carrying capability. (3) Irritants in smoke cause increased mucus secretion by the mucosa of the bronchial tree and swelling of the mucosal lining, both of which impede airflow into and out of the lungs. (4) Irritants in smoke also inhibit the movement of cilia and destroy cilia in the lining of the respiratory system. Thus, excess mucus and foreign debris are not easily removed, which further adds to the difficulty in breathing. (5) With time, smoking leads to destruction of elastic fibers in the lungs and is the prime cause of emphysema (described on page 913). These changes cause collapse of small bronchioles and trapping of air in alveoli at the end of exhalation. The result is less efficient gas exchange. ■

Medical Terminology

Asphyxia (as-FIK-sē-a; *sphyxia* = pulse) Oxygen starvation due to low atmospheric oxygen or interference with ventilation, external respiration, or internal respiration.

Aspiration (as′-pi-RĀ-shun) Inhalation of a foreign substance such as water, food, or a foreign body into the bronchial tree; also, the drawing of a substance in or out by suction.

Black lung disease A condition in which the lungs appear black instead of pink due to inhalation of coal dust over a period of many years. Most often it affects people who work in the coal industry.

Bronchiectasis (bron′-kē-EK-ta-sis; *-ektasis* = stretching) A chronic dilation of the bronchi or bronchioles resulting from damage to the bronchial wall, for example, from respiratory infections.

Cheyne–Stokes respiration (CHĀN STŌKS res′-pi-RĀ-shun) A repeated cycle of irregular breathing that begins with shallow breaths that increase in depth and rapidity and then decrease and cease altogether for 15 to 20 seconds. Cheyne–Stokes is normal in infants; it is also often seen just before death from pulmonary, cerebral, cardiac, and kidney disease.

Dyspnea (DISP-nē-a; *dys-* = painful, difficult) Painful or labored breathing.

Epistaxis (ep′-i-STAK-sis) Loss of blood from the nose due to trauma, infection, allergy, malignant growths, or bleeding disorders. It can be arrested by cautery with silver nitrate, electrocautery, or firm packing. Also called **nosebleed.**

Hypoventilation (*hypo-* = below) Slow and shallow breathing.

Mechanical ventilation The use of an automatically cycling device (ventilator or respirator) to assist breathing. A plastic tube is inserted into the nose or mouth and the tube is attached to a device that forces air into the lungs. Exhalation occurs passively due to the elastic recoil of the lungs.

Rales (RĀLS) Sounds sometimes heard in the lungs that resemble bubbling or rattling. Rales are to the lungs what murmurs are to the heart. Different types are due to the presence of an abnormal type or amount of fluid or mucus within the bronchi or alveoli, or to bronchoconstriction that causes turbulent airflow.

Respirator (RES-pi-rā′-tor) An apparatus fitted to a mask over the nose and mouth, or hooked directly to an endotracheal or tracheotomy tube, that is used to assist or support ventilation or to provide nebulized medication to the air passages.

Respiratory failure A condition in which the respiratory system either cannot supply sufficient O_2 to maintain metabolism or cannot eliminate enough CO_2 to prevent respiratory acidosis (a lower-than-normal pH in interstitial fluid).

Rhinitis (rī-NĪ-tis; *rhin-* = nose) Chronic or acute inflammation of the mucous membrane of the nose due to viruses, bacteria, or irritants. Excessive mucus production produces a runny nose, nasal congestion, and postnasal drip.

Sleep apnea (AP-nē-a; *a-* = without; *-pnea* = breath) A disorder in which a person repeatedly stops breathing for 10 or more seconds while sleeping. Most often, it occurs because loss of muscle tone in pharyngeal muscles allows the airway to collapse.

Sputum (SPŪ-tum = to spit) Mucus and other fluids from the air passages that is expectorated (expelled by coughing).

Strep throat Inflammation of the pharynx caused by the bacterium *Streptococcus pyogenes.* It may also involve the tonsils and middle ear.

Tachypnea (tak′-ip-NĒ-a; *tachy-* = rapid; *-pnea* = breath) Rapid breathing rate.

Wheeze (HWEĒZ) A whistling, squeaking, or musical high-pitched sound during breathing resulting from a partially obstructed airway.

Chapter 12

The Digestive System

The **digestive system** breaks down food into forms that can be absorbed and used by body cells. It also absorbs water, vitamins, and minerals, and eliminates wastes from the body. This tubular system which extends from the mouth to the anus includes the oral cavity, esophagus, stomach, pancreas, liver, gallbladder, small intestine, and large intestine. The medical specialty that deals with the structure, function, diagnosis, and treatment of diseases of the stomach and intestines is called *gastroenterology*. The medical specialty that deals with the diagnosis and treatment of disorders of the rectum and anus is called *proctology*.

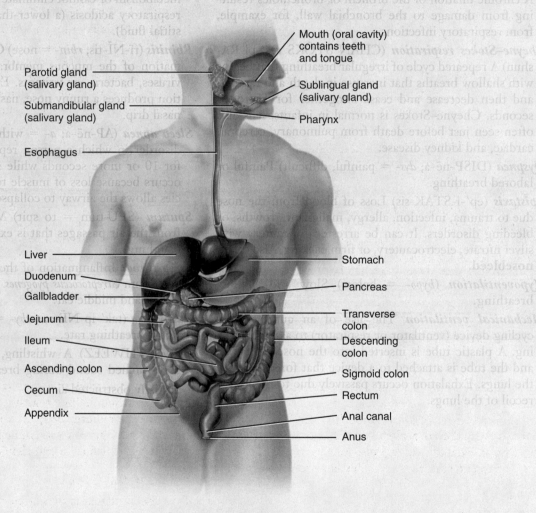

Parotid gland (salivary gland)

Submandibular gland (salivary gland)

Esophagus

Liver

Duodenum

Gallbladder

Jejunum

Ileum

Ascending colon

Cecum

Appendix

Mouth (oral cavity) contains teeth and tongue

Sublingual gland (salivary gland)

Pharynx

Stomach

Pancreas

Transverse colon

Descending colon

Sigmoid colon

Rectum

Anal canal

Anus

Organs of the digestive system

Functions

1. Ingestion: taking food into the mouth.
2. Secretion: release of water, acid, buffers, and enzymes into the lumen of the GI tract.
3. Mixing and propulsion: churning and propulsion of food through the GI tract.
4. Digestion: mechanical and chemical breakdown of food.
5. Absorption: passage of digested products from the GI tract into the blood and lymph.
6. Defecation: the elimination of feces from the GI tract.

Tests, Procedures, and Techniques

■ Colonoscopy

Colonoscopy (kō-lon-OS-kō-pē; *-skopes* = to view) is the visual examination of the lining of the colon using an elongated, flexible, fiberoptic endoscope called a *colonoscope*. It is used to detect disorders such as polyps, cancer, and diverticulosis, to take tissue samples, and to remove small polyps. Most tumors of the large intestine occur in the rectum.

■ Colostomy

Colostomy (kō-LOS-tō-mē; *-stomy* = provide an opening) is the diversion of feces through an opening in the colon, creating a surgical "stoma" (artificial opening) that is made in the exterior of the abdominal wall. This opening serves as a substitute anus through which feces are eliminated into a bag worn on the abdomen.

■ Gastroscopy

Gastroscopy (gas-TROS-kō-pē; *-scopy* = to view with a lighted instrument) is the endoscopic examination of the stomach in which the examiner can view the interior of the stomach directly to evaluate an ulcer, tumor, inflammation, or source of bleeding.

Occult Blood

The term **occult blood** refers to blood that is hidden; it is not detectable by the human eye. The main diagnostic value of occult blood testing is to screen for colorectal cancer. Two substances often examined for occult blood are feces and urine. Several types of products are available for at-home testing for hidden blood in feces. The tests are based on color changes when reagents are added to feces. The presence of occult blood in urine may be detected at home by using dip-and-read reagent strips.

■ Root Canal Therapy

Root canal therapy is a multistep procedure in which all traces of pulp tissue are removed from the pulp cavity and root canals of a badly diseased tooth. After a hole is made in the tooth, the root canals are filed out and irrigated to remove bacteria. Then, the canals are treated with medication and sealed tightly. The damaged crown is then repaired. ■

Disorders Affecting the Digestive System

■ Anorexia Nervosa

Anorexia nervosa is a chronic disorder characterized by self-induced weight loss, negative perception of body image, and physiological changes that result from nutritional depletion. Patients with anorexia nervosa have a fixation on weight control and often insist on having a bowel movement every day despite inadequate food intake. They often abuse laxatives, which worsens the fluid and electrolyte imbalances and nutrient deficiencies. The disorder is found predominantly in young, single females, and it may be inherited. Abnormal patterns of menstruation, amenorrhea (absence of menstruation), and a lowered basal metabolic rate reflect the depressant effects of starvation. Individuals may become emaciated and may ultimately die of starvation or one of its complications. Also associated with the disorder are osteoporosis, depression, and brain abnormalities coupled with impaired mental performance. Treatment consists of psychotherapy and dietary regulation.

■ Appendicitis

Inflammation of the appendix, termed **appendicitis,** is preceded by obstruction of the lumen of the appendix by chyme, inflammation, a foreign body, a carcinoma of the cecum, stenosis, or kinking of the organ. It is characterized by high fever, elevated white blood cell count, and a neutrophil count higher than 75%. The infection that follows may result in edema and ischemia and may progress to gangrene and perforation within 24 hours. Typically, appendicitis begins with referred pain in the umbilical region of the abdomen, followed by anorexia (loss of appetite for food), nausea, and vomiting. After several hours the pain localizes in the right lower quadrant (RLQ) and is continuous, dull or severe, and intensified by coughing, sneezing, or body movements. Early appendectomy (removal of the appendix) is recommended because it is safer to operate than to risk rupture, peritonitis, and gangrene. Although it required major abdominal surgery in the past, today appendectomies are usually performed laparoscopically.

■ Bulimia

Bulimia (*bu-* = ox; *limia* = hunger) or *binge–purge syndrome* is a disorder that typically affects young, single, middle-class, white females, characterized by overeating at least twice a week followed by purging by self-induced vomiting, strict dieting or fasting, vigorous exercise, or use of laxatives or diuretics; it occurs in response to fears of being overweight or to stress, depression, and physiological disorders such as hypothalamic tumors.

■ Colorectal Cancer

Colorectal cancer is among the deadliest of malignancies, ranking second to lung cancer in males and third after lung cancer and breast cancer in females. Genetics plays a very important role; an inherited predisposition contributes to more than half of all cases of colorectal cancer. Intake of alcohol and diets high in animal fat and protein are associated with increased risk of colorectal cancer; dietary fiber, retinoids, calcium, and selenium may be protective. Signs and symptoms of colorectal cancer include diarrhea, constipation, cramping, abdominal pain, and rectal bleeding, either visible or occult (hidden in feces). Precancerous growths on the mucosal surface, called **polyps,** also increase the risk of developing colorectal cancer.

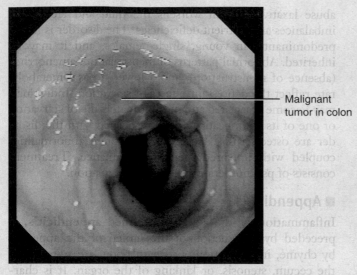

Malignant tumor in colon

Screening for colorectal cancer includes testing for blood in the feces, digital rectal examination, sigmoidoscopy, colonoscopy, and barium enema. Tumors may be removed endoscopically or surgically.

Diverticular Disease

In **diverticular disease,** saclike outpouchings of the wall of the colon, termed **diverticula,** occur in places where the muscularis has weakened and may become inflamed. Development of diverticula is known as **diverticulosis.** Many people who develop diverticulosis have no symptoms and experience no complications. Of those people known to have diverticulosis, 10–25% eventually develop an inflammation known as **diverticulitis.** This condition may be characterized by pain, either constipation or increased frequency of defecation, nausea, vomiting, and low-grade fever. Because diets low in fiber contribute to development of diverticulitis, patients who change to high-fiber diets show marked relief of symptoms. In severe cases, affected portions of the colon may require surgical removal. If diverticula rupture, the release of bacteria into the abdominal cavity can cause peritonitis.

Gastroenteritis

Gastroenteritis (gas'-trō-en-ter-Ī-tis; *gastro-* = stomach; *enteron* = intestine; *-itis* = inflammation) is the inflammation of the lining of the stomach and intestine (especially the small intestine). It is usually caused by a viral or bacterial infection that may be acquired by contaminated food or water or by people in close contact. Symptoms include diarrhea, vomiting, fever, loss of appetite, cramps, and abdominal discomfort.

Gastroesophageal Reflux Disease

If the lower esophageal sphincter fails to close adequately after food has entered the stomach, the stomach contents can reflux (back up) into the inferior portion of the esophagus. This condition is known as **gastroesophageal reflux disease (GERD).** Hydrochloric acid (HCl) from the stomach contents can irritate the esophageal wall, resulting in a burning sensation that is called **heartburn** because it is experienced in a region very near the heart; it is unrelated to any cardiac problem. Drinking alcohol and smoking can cause the sphincter to relax, worsening the problem. The symptoms of GERD often can be controlled by avoiding foods that strongly stimulate stomach acid secretion (coffee, chocolate, tomatoes, fatty foods, orange juice, peppermint, spearmint, and onions). Other acid-reducing strategies include taking over-the-counter histamine-2 (H_2) blockers such as Tagamet HB® or Pepcid AC® 30 to 60 minutes before eating to block acid secretion, and neutralizing acid that has already been secreted with antacids such as Tums® or Maalox®. Symptoms are less likely to occur if food is eaten in smaller amounts and if the person does not lie down immediately after a meal. GERD may be associated with cancer of the esophagus.

Hepatitis

Hepatitis is an inflammation of the liver that can be caused by viruses, drugs, and chemicals, including alcohol. Clinically, several types of viral hepatitis are recognized.

Hepatitis A (infectious hepatitis) is caused by the hepatitis A virus and is spread via fecal contamination of objects such as food, clothing, toys, and eating utensils (fecal–oral route). It is generally a mild disease of children and young adults characterized by loss of appetite, malaise, nausea, diarrhea, fever, and chills. Eventually, jaundice appears. This type of hepatitis does not cause lasting liver damage. Most people recover in 4 to 6 weeks.

Hepatitis B is caused by the hepatitis B virus and is spread primarily by sexual contact and contaminated syringes and transfusion equipment. It can also be spread via saliva and tears. Hepatitis B virus can be present for years or even a lifetime, and it can produce cirrhosis and possibly cancer of the liver. Individuals who harbor the active hepatitis B virus also become carriers. Vaccines produced through recombinant DNA technology (for example, Recombivax HB®) are available to prevent hepatitis B infection.

Hepatitis C, caused by the hepatitis C virus, is clinically similar to hepatitis B. Hepatitis C can cause cirrhosis and possibly liver cancer. In developed nations, donated blood is screened for the presence of hepatitis B and C.

Hepatitis D is caused by the hepatitis D virus. It is transmitted like hepatitis B and in fact a person must have been co-infected with hepatitis B before contracting hepatitis D. Hepatitis D results in severe liver damage and has a higher fatality rate than infection with hepatitis B virus alone.

Hepatitis E is caused by the hepatitis E virus and is spread like hepatitis A. Although it does not cause chronic liver disease, hepatitis E virus has a very high mortality rate among pregnant women.

Inflammatory bowel disease

Inflammatory bowel disease (in-FLAM-a-tō'-rē BOW-el) is inflammation of the gastrointestinal tract that exists in two forms. (1) *Crohn's disease* is an inflammation of any part of the gastrointestinal tract in which the inflammation extends from the mucosa through the submucosa, muscularis, and serosa. (2) *Ulcerative colitis* is an inflammation of the mucosa of the colon and rectum, usually accompanied by rectal bleeding. Curiously, cigarette smoking increases the risk of Crohn's disease but decreases the risk of ulcerative colitis.

■ Irritable bowel syndrome

Irritable bowel syndrome (IBS) is a disease of the entire gastrointestinal tract in which a person reacts to stress by developing symptoms (such as cramping and abdominal pain) associated with alternating patterns of diarrhea and constipation. Excessive amounts of mucus may appear in feces; other symptoms include flatulence, nausea, and loss of appetite. The condition is also known as **irritable colon** or **spastic colitis.**

■ Mumps

Although any of the salivary glands may be the target of a nasopharyngeal infection, the mumps virus (*paramyxovirus*) typically attacks the parotid glands. **Mumps** is an inflammation and enlargement of the parotid glands accompanied by moderate fever, malaise (general discomfort), and extreme pain in the throat, especially when swallowing sour foods or acidic juices. Swelling occurs on one or both sides of the face, just anterior to the ramus of the mandible. In about 30% of males past puberty, the testes may also become inflamed; sterility rarely occurs because testicular involvement is usually unilateral (one testis only). Since a vaccine became available for mumps in 1967, the incidence of the disease has declined dramatically.

■ Pancreatitis and Pancreatic Cancer

Inflammation of the pancreas, as may occur in association with alcohol abuse or chronic gallstones, is called **pancreatitis** (pan′-krē-a-TĪ-tis). In a more severe condition known as **acute pancreatitis,** which is associated with heavy alcohol intake or biliary tract obstruction, the pancreatic cells may release either trypsin instead of trypsinogen or insufficient amounts of trypsin inhibitor, and the trypsin begins to digest the pancreatic cells. Patients with acute pancreatitis usually respond to treatment, but recurrent attacks are the rule. In some people pancreatitis is idiopathic, meaning that the cause is unknown. Other causes of pancreatitis include cystic fibrosis, high levels of calcium in the blood (hypercalcemia), high levels of blood fats (hyperlipidemia or hypertriglyceridemia), some drugs, and certain autoimmune conditions. However, in roughly 70% of adults with pancreatitis, the cause is alcoholism. Often the first episode happens between ages 30 and 40.

Pancreatic cancer usually affects people over 50 years of age and occurs more frequently in males. Typically, there are few symptoms until the disorder reaches an advanced stage and often not until it has metastasized to other parts of the body such as the lymph nodes, liver, or lungs. The disease is nearly always fatal and is the fourth most common cause of death from cancer in the United States. Pancreatic cancer has been linked to fatty foods, high alcohol consumption, genetic factors, smoking, and chronic pancreatitis.

■ Peptic Ulcer Disease

In the United States, 5–10% of the population develops **peptic ulcer disease (PUD).** An **ulcer** is a craterlike lesion in a membrane; ulcers that develop in areas of the GI tract exposed to acidic gastric juice are called **peptic ulcers.** The most common complication of peptic ulcers is bleeding, which can lead to anemia if enough blood is lost. In acute cases, peptic ulcers can lead to shock and death. Three distinct causes of PUD are recognized: (1) the bacterium *Helicobacter pylori*; (2) nonsteroidal anti-inflammatory drugs (NSAIDs) such as aspirin; and (3) hypersecretion of HCl, as occurs in Zollinger–Ellison syndrome, a gastrin-producing tumor, usually of the pancreas.

Helicobacter pylori (previously named *Campylobacter pylori*) is the most frequent cause of PUD. The bacterium produces an enzyme called urease, which splits urea into ammonia and carbon dioxide. While shielding the bacterium from the acidity of the stomach, the ammonia also damages the protective mucous layer of the stomach and the underlying gastric cells. *H. pylori* also produces catalase, an enzyme that may protect the microbe from phagocytosis by neutrophils, plus several adhesion proteins that allow the bacterium to attach itself to gastric cells.

Several therapeutic approaches are helpful in the treatment of PUD. Cigarette smoke, alcohol, caffeine, and NSAIDs should be avoided because they can impair mucosal defensive mechanisms, which increases mucosal susceptibility to the damaging effects of HCl. In cases associated with *H. pylori*, treatment with an antibiotic drug often resolves the problem. Oral antacids such as Tums® or Maalox® can help temporarily by buffering gastric acid.

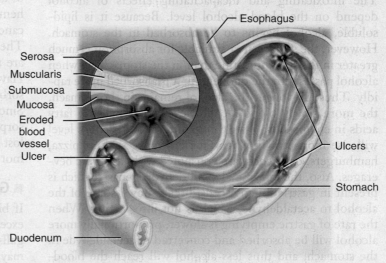

Peptic ulcer

When hypersecretion of HCl is the cause of PUD, H$_2$ blockers (such as Tagamet®) or proton pump inhibitors such as omeprazole (Prilosec®), which block secretion of H$^+$ from parietal cells, may be used.

Periodontal Disease

Periodontal disease is a collective term for a variety of conditions characterized by inflammation and degeneration of the gingivae, alveolar bone, periodontal ligament, and cementum. In one such condition, called **pyorrhea,** initial symptoms include enlargement and inflammation of the soft tissue and bleeding of the gums. Without treatment, the soft tissue may deteriorate and the alveolar bone may be resorbed, causing loosening of the teeth and recession of the gums. Periodontal diseases are often caused by poor oral hygiene; by local irritants, such as bacteria, impacted food, and cigarette smoke; or by a poor "bite."

Peritonitis

A common cause of **peritonitis,** an acute inflammation of the peritoneum, is contamination of the peritoneum by infectious microbes, which can result from accidental or surgical wounds in the abdominal wall, or from perforation or rupture of abdominal organs. If, for example, bacteria gain access to the peritoneal cavity through an intestinal perforation or rupture of the appendix, they can produce an acute, life-threatening form of peritonitis. A less serious (but still painful) form of peritonitis can result from the rubbing together of inflamed peritoneal surfaces. Peritonitis is of particularly grave concern to those who rely on peritoneal dialysis, a procedure in which the peritoneum is used to filter the blood when the kidneys do not function properly.

Additional Clinical Considerations

Absorption of Alcohol

The intoxicating and incapacitating effects of alcohol depend on the blood alcohol level. Because it is lipid-soluble, alcohol begins to be absorbed in the stomach. However, the surface area available for absorption is much greater in the small intestine than in the stomach, so when alcohol passes into the duodenum, it is absorbed more rapidly. Thus, the longer the alcohol remains in the stomach, the more slowly blood alcohol level rises. Because fatty acids in chyme slow gastric emptying, blood alcohol level will rise more slowly when fat-rich foods, such as pizza, hamburgers, or nachos, are consumed with alcoholic beverages. Also, the enzyme alcohol dehydrogenase, which is present in gastric mucosa cells, breaks down some of the alcohol to acetaldehyde, which is not intoxicating. When the rate of gastric emptying is slower, proportionally more alcohol will be absorbed and converted to acetaldehyde in the stomach, and thus less alcohol will reach the bloodstream. Given identical consumption of alcohol, females often develop higher blood alcohol levels (and therefore experience greater intoxication) than males of comparable size because the activity of gastric alcohol dehydrogenase is up to 60% lower in females than in males. Asian males may also have lower levels of this gastric enzyme.

Dental Caries

Dental caries, or tooth decay, involves a gradual demineralization (softening) of the enamel and dentin. If untreated, microorganisms may invade the pulp, causing inflammation and infection, with subsequent death of the pulp and abscess of the alveolar bone surrounding the root's apex, requiring root canal therapy.

Dental caries begin when bacteria, acting on sugars, produce acids that demineralize the enamel. **Dextran,** a sticky polysaccharide produced from sucrose, causes the bacteria to stick to the teeth. Masses of bacterial cells, dextran, and other debris adhering to teeth constitute **dental plaque.** Saliva cannot reach the tooth surface to buffer the acid because the plaque covers the teeth. Brushing the teeth after eating removes the plaque from flat surfaces before the bacteria can produce acids. Dentists also recommend that the plaque between the teeth be removed every 24 hours with dental floss.

Dietary Fiber

Dietary fiber consists of indigestible plant carbohydrates—such as cellulose, lignin, and pectin—found in fruits, vegetables, grains, and beans. **Insoluble fiber,** which does not dissolve in water, includes the woody or structural parts of plants such as the skins of fruits and vegetables and the bran coating around wheat and corn kernels. Insoluble fiber passes through the GI tract largely unchanged but speeds up the passage of material through the tract. **Soluble fiber,** which does dissolve in water, forms a gel that slows the passage of material through the tract. It is found in abundance in beans, oats, barley, broccoli, prunes, apples, and citrus fruits.

People who choose a fiber-rich diet may reduce their risk of developing obesity, diabetes, atherosclerosis, gallstones, hemorrhoids, diverticulitis, appendicitis, and colorectal cancer. Soluble fiber also may help lower blood cholesterol. The liver normally converts cholesterol to bile salts, which are released into the small intestine to help fat digestion. Having accomplished their task, the bile salts are reabsorbed by the small intestine and recycled back to the liver. Since soluble fiber binds to bile salts to prevent their reabsorption, the liver makes more bile salts to replace those lost in feces. Thus, the liver uses more cholesterol to make more bile salts and blood cholesterol level is lowered.

Gallstones

If bile contains either insufficient bile salts or lecithin or excessive cholesterol, the cholesterol may crystallize to form **gallstones.** As they grow in size and number, gallstones may cause minimal, intermittent, or complete obstruction to the flow of bile from the gallbladder into the duodenum.

Treatment consists of using gallstone-dissolving drugs, lithotripsy (shock-wave therapy), or surgery. For people with recurrent gallstones or for whom drugs or lithotripsy is not indicated, *cholecystectomy*—the removal of the gallbladder and its contents—is necessary. More than half a million cholecystectomies are performed each year in the United States.

■ Hemorrhoids

Hemorrhoids (HEM-ō-royds; *hemi* = blood; *rhoia* = flow) are varicosed (enlarged and inflamed) superior rectal veins. Hemorrhoids develop when the veins are put under pressure and become engorged with blood. If the pressure continues, the wall of the vein stretches. Such a distended vessel oozes blood; bleeding or itching is usually the first sign that a hemorrhoid has developed. Stretching of a vein also favors clot formation, further aggravating swelling and pain. Hemorrhoids may be caused by constipation, which may be brought on by low-fiber diets. Also, repeated straining during defecation forces blood down into the rectal veins, increasing pressure in those veins and possibly causing hemorrhoids. Also called **piles.**

■ Jaundice

Jaundice (JAWN-dis = yellowed) is a yellowish coloration of the sclerae (whites of the eyes), skin, and mucous membranes due to a buildup of a yellow compound called bilirubin. After bilirubin is formed from the breakdown of the heme pigment in aged red blood cells, it is transported to the liver, where it is processed and eventually excreted into bile. The three main categories of jaundice are (1) *prehepatic jaundice*, due to excess production of bilirubin; (2) *hepatic jaundice*, due to congenital liver disease, cirrhosis of the liver, or hepatitis; and (3) *extrahepatic jaundice*, due to blockage of bile drainage by gallstones or cancer of the bowel or the pancreas.

Because the liver of a newborn functions poorly for the first week or so, many babies experience a mild form of jaundice called *neonatal (physiological) jaundice* that disappears as the liver matures. Usually, it is treated by exposing the infant to blue light, which converts bilirubin into substances the kidneys can excrete.

■ Lactose Intolerance

In some people the absorptive cells of the small intestine fail to produce enough lactase, which, as you just learned, is essential for the digestion of lactose. This results in a condition called **lactose intolerance,** in which undigested lactose in chyme causes fluid to be retained in the feces; bacterial fermentation of the undigested lactose results in the production of gases. Symptoms of lactose intolerance include diarrhea, gas, bloating, and abdominal cramps after consumption of milk and other dairy products. The symptoms can be relatively minor or serious enough to require medical attention. The *hydrogen breath test* is often used to aid in diagnosis of lactose intolerance. Very little hydrogen can be detected in the breath of a normal person, but hydrogen is among the gases produced when undigested lactose in the colon is fermented by bacteria. The hydrogen is absorbed from the intestines and carried through the bloodstream to the lungs, where it is exhaled. Persons with lactose intolerance can take dietary supplements to aid in the digestion of lactose.

■ Polyps in the Colon

Polyps in the colon are generally slow-developing benign growths that arise from the mucosa of the large intestine. Often, they do not cause symptoms. If symptoms do occur, they include diarrhea, blood in the feces, and mucus discharged from the anus. The polyps are removed by colonoscopy or surgery because some of them may become cancerous.

■ Pylorospasm and Pyloric Stenosis

Two abnormalities of the pyloric sphincter can occur in infants. In **pylorospasm** (pī-LOR-ō-spazm), the smooth muscle fibers of the sphincter fail to relax normally, so food does not pass easily from the stomach to the small intestine, the stomach becomes overly full, and the infant vomits often to relieve the pressure. Pylorospasm is treated by drugs that relax the muscle fibers of the pyloric sphincter. **Pyloric stenosis** (ste-NŌ-sis) is a narrowing of the pyloric sphincter that must be corrected surgically. The hallmark symptom is *projectile vomiting*—the spraying of liquid vomitus some distance from the infant.

■ Vomiting

Vomiting or *emesis* is the forcible expulsion of the contents of the upper GI tract (stomach and sometimes duodenum) through the mouth. The strongest stimuli for vomiting are irritation and distension of the stomach; other stimuli include unpleasant sights, general anesthesia, dizziness, and certain drugs such as morphine and derivatives of digitalis. Nerve impulses are transmitted to the vomiting center in the medulla oblongata, and returning impulses propagate to the upper GI tract organs, diaphragm, and abdominal muscles. Vomiting involves squeezing the stomach between the diaphragm and abdominal muscles and expelling the contents through open esophageal sphincters. Prolonged vomiting, especially in infants and elderly people, can be serious because the loss of acidic gastric juice can lead to alkalosis (higher than normal blood pH), dehydration, and damage to the esophagus and teeth.

Medical Terminology

Achalasia (ak′-a-LĀ-zē-a; *a-* = without; *chalasis* = relaxation) A condition caused by malfunction of the myenteric plexus in which the lower esophageal sphincter fails to relax normally as food approaches. A whole meal may become lodged in the esophagus and enter the stomach very slowly. Distension of the esophagus results in chest pain that is often confused with pain originating from the heart.

Borborygmus (bor'-bō-RIG-mus) A rumbling noise caused by the propulsion of gas through the intestines.

Canker sore (KANG-ker) Painful ulcer on the mucous membrane of the mouth that affects females more often than males, usually between ages 10 and 40; may be an autoimmune reaction or a food allergy.

Cirrhosis Distorted or scarred liver as a result of chronic inflammation due to hepatitis, chemicals that destroy hepatocytes, parasites that infect the liver, or alcoholism; the hepatocytes are replaced by fibrous or adipose connective tissue. Symptoms include jaundice, edema in the legs, uncontrolled bleeding, and increased sensitivity to drugs.

Colitis (ko-LĪ-tis) Inflammation of the mucosa of the colon and rectum in which absorption of water and salts is reduced, producing watery, bloody feces and, in severe cases, dehydration and salt depletion. Spasms of the irritated muscularis produce cramps. It is thought to be an autoimmune condition.

Dysphagia (dis-FĀ-jē-a; *dys-* = abnormal; *phagia* = to eat) Difficulty in swallowing that may be caused by inflammation, paralysis, obstruction, or trauma.

Flatus (FLĀ-tus) Air (gas) in the stomach or intestine, usually expelled through the anus. If the gas is expelled through the mouth, it is called **eructation** or **belching** (burping). Flatus may result from gas released during the breakdown of foods in the stomach or from swallowing air or gas-containing substances such as carbonated drinks.

Food poisoning A sudden illness caused by ingesting food or drink contaminated by an infectious microbe (bacterium, virus, or protozoan) or a toxin (poison). The most common cause of food poisoning is the toxin produced by the bacterium *Staphylococcus aureus*. Most types of food poisoning cause diarrhea and/or vomiting, often associated with abdominal pain.

Halitosis (hal'-ī-TŌ-sis; *halitus-* = breath; *-osis* = condition) A foul odor from the mouth; also called **bad breath.**

Heartburn A burning sensation in a region near the heart due to irritation of the mucosa of the esophagus from hydrochloric acid in stomach contents. It is caused by failure of the lower esophageal sphincter to close properly, so that the stomach contents enter the inferior esophagus. It is not related to any cardiac problem.

Hernia (HER-nē-a) Protrusion of all or part of an organ through a membrane or cavity wall, usually the abdominal cavity. *Hiatus (diaphragmatic) hernia* is the protrusion of a part of the stomach into the thoracic cavity through the esophageal hiatus of the diaphragm. *Inguinal hernia* is the protrusion of the hernial sac into the inguinal opening; it may contain a portion of the bowel in an advanced stage and may extend into the scrotal compartment in males, causing strangulation of the herniated part.

Malabsorption (mal-ab-SORP-shun; *mal-* = bad) A number of disorders in which nutrients from food are not absorbed properly. It may be due to disorders that result in the inadequate breakdown of food during digestion (due to inadequate digestive enzymes or juices), damage to the lining of the small intestine (from surgery, infections, and drugs like neomycin and alcohol), and impairment of motility. Symptoms may include diarrhea, weight loss, weakness, vitamin deficiencies, and bone demineralization.

Malocclusion (mal'-ō-KLOO-zhun; *mal-* = bad; *occlusion* = to fit together) Condition in which the surfaces of the maxillary (upper) and mandibular (lower) teeth fit together poorly.

Nausea (NAW-sē-a; *nausia* = seasickness) Discomfort characterized by a loss of appetite and the sensation of impending vomiting. Its causes include local irritation of the gastrointestinal tract, a systemic disease, brain disease or injury, overexertion, or the effects of medication or drug overdosage.

Traveler's diarrhea Infectious disease of the gastrointestinal tract that results in loose, urgent bowel movements, cramping, abdominal pain, malaise, nausea, and occasionally fever and dehydration. It is acquired through ingestion of food or water contaminated with fecal material typically containing bacteria (especially *Escherichia coli*); viruses or protozoan parasites are less common causes.

Chapter 13 | The Urinary System

The **urinary system** consists of two kidneys, two ureters, one urinary bladder, and one urethra. It contributes to the homeostasis of the body by altering blood composition, pH, volume, and pressure. It maintains blood osmolarity, and excretes wastes and foreign substances. The urinary system also produces some hormones. *Nephrology* is the scientific study of the anatomy, physiology, and pathology of the kidneys. The branch of medicine that deals with the male and female urinary systems and the male reproductive system is called *urology*.

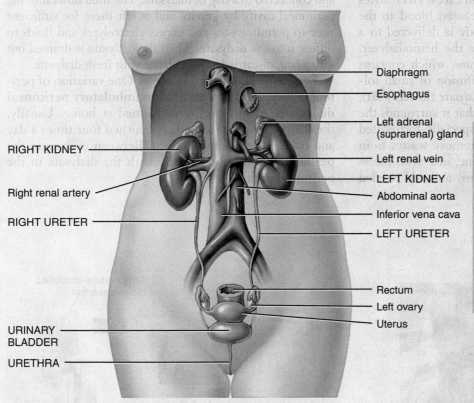

- Diaphragm
- Esophagus
- Left adrenal (suprarenal) gland
- Left renal vein
- LEFT KIDNEY
- Abdominal aorta
- Inferior vena cava
- LEFT URETER
- Rectum
- Left ovary
- Uterus

RIGHT KIDNEY
Right renal artery
RIGHT URETER
URINARY BLADDER
URETHRA

Organs of the urinary system in a female

Functions of the Urinary System

1. The kidneys regulate blood volume and composition, help regulate blood pressure, synthesize glucose, release erythropoietin, participate in vitamin D synthesis, and excrete wastes by forming urine.
2. The ureters transport urine from the kidneys to the urinary bladder.
3. The urinary bladder stores urine.
4. The urethra discharges urine from the body.

Tests, Procedures, and Techniques

■ Cystoscopy

Cystoscopy (sis-TOS-kō-pē; *cysto-* = bladder; *-skopy* = to examine) is a very important procedure for direct examination of the mucosa of the urethra and urinary bladder and prostate in males. In the procedure, a *cystoscope* (a flexible narrow tube with a light) is inserted into the urethra to examine the structures through which it passes. With special attachments, tissue samples can be removed for examination (biopsy) and small stones can be removed. Cystoscopy is useful for evaluating urinary bladder problems such as cancer and infections. It can also evaluate the degree of obstruction resulting from an enlarged prostate.

■ Diuretics

Diuretics are substances that slow renal reabsorption of water and thereby cause *diuresis*, an elevated urine flow rate, which in turn reduces blood volume. Diuretic drugs often are prescribed to treat *hypertension* (high blood pressure) because lowering blood volume usually reduces blood pressure. Naturally occurring diuretics include *caffeine* in coffee, tea, and sodas, which inhibits Na^+ reabsorption, and *alcohol* in beer, wine, and mixed drinks, which inhibits secretion of ADH. Most diuretic drugs act by interfering with a mechanism for reabsorption of filtered Na^+. For example, loop diuretics, such as furosemide (Lasix®), selectively inhibit the Na^+–K^+–$2Cl^-$ symporters in the thick ascending limb of the loop of Henle. The thiazide diuretics, such as chlorthiazide (Diuril®), act in the distal convoluted tubule, where they promote loss of Na^+ and Cl^- in the urine by inhibiting Na^+–Cl^- symporters.

■ Dialysis

If a person's kidneys are so impaired by disease or injury that he or she is unable to function adequately, then blood must be cleansed artificially by **dialysis** (dī-AL-i-sis; *dialyo* = to separate), the separation of large solutes from smaller ones by diffusion through a selectively permeable membrane. One method of dialysis is **hemodialysis** (hē-mō-dī-AL-i-sis; *hemo-* = blood), which directly filters the patient's blood by removing wastes and excess electrolytes and fluid and then returning the cleansed blood to the patient. Blood removed from the body is delivered to a *hemodialyzer* (artificial kidney). Inside the hemodialyzer, blood flows through a *dialysis membrane*, which contains pores large enough to permit the diffusion of small solutes. A special solution, called the *dialysate* (dī-AL-i-sāt), is pumped into the hemodialyzer so that it surrounds the dialysis membrane. The dialysate is specially formulated to maintain diffusion gradients that remove wastes from the blood (for example, urea, creatinine, uric acid, excess phosphate, potassium, and sulfate ions) and add needed substances (for example, glucose and bicarbonate ions) to it. The cleansed blood is passed through an air embolus detector to remove air and then returned to the body. An anticoagulant (heparin) is added to prevent blood from clotting in the hemodialyzer. As a rule, most people on hemodialysis require about 6–12 hours a week, typically divided into three sessions.

Another method of dialysis, called **peritoneal dialysis,** uses the peritoneum of the abdominal cavity as the dialysis membrane to filter the blood. The peritoneum has a large surface area and numerous blood vessels, and is a very effective filter. A catheter is inserted into the peritoneal cavity and connected to a bag of dialysate. The fluid flows into the peritoneal cavity by gravity and is left there for sufficient time to permit wastes and excess electrolytes and fluids to diffuse into the dialysate. Then the dialysate is drained out into a bag, discarded, and replaced with fresh dialysate.

Each cycle is called an *exchange*. One variation of peritoneal dialysis, called **continuous ambulatory peritoneal dialysis (CAPD),** can be performed at home. Usually, the dialysate is drained and replenished four times a day and once at night during sleep. Between exchanges the person can move about freely with the dialysate in the peritoneal cavity.

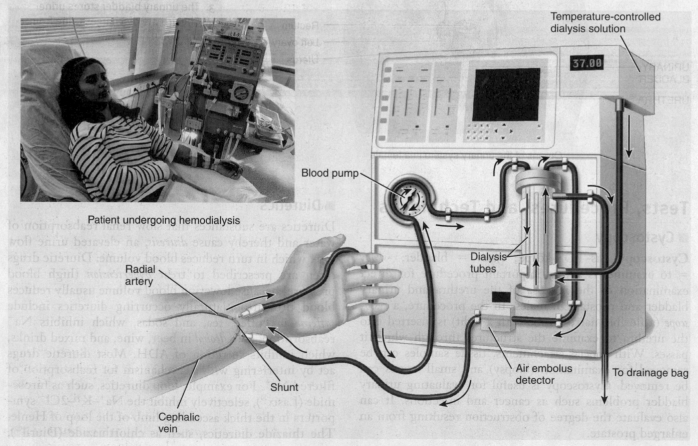

Patient undergoing hemodialysis

Temperature-controlled dialysis solution

Blood pump

Radial artery

Shunt

Cephalic vein

Dialysis tubes

Air embolus detector

To drainage bag

Dialysis

Disorders Affecting the Urinary System

■ Glomerular Diseases

A variety of conditions may damage the kidney glomeruli, either directly or indirectly because of disease elsewhere in the body. Typically, the filtration membrane sustains damage, and its permeability increases.

Glomerulonephritis is an inflammation of the kidney that involves the glomeruli. One of the most common causes is an allergic reaction to the toxins produced by streptococcal bacteria that have recently infected another part of the body, especially the throat. The glomeruli become so inflamed, swollen, and engorged with blood that the filtration membranes allow blood cells and plasma proteins to enter the filtrate. As a result, the urine contains many erythrocytes (hematuria) and a lot of protein. The glomeruli may be permanently damaged, leading to chronic renal failure.

Nephrotic syndrome is a condition characterized by *proteinuria* (protein in the urine) and *hyperlipidemia* (high blood levels of cholesterol, phospholipids, and triglycerides). The proteinuria is due to an increased permeability of the filtration membrane, which permits proteins, especially albumin, to escape from blood into urine. Loss of albumin results in *hypoalbuminemia* (low blood albumin level) once liver production of albumin fails to meet increased urinary losses. Edema, usually seen around the eyes, ankles, feet, and abdomen, occurs in nephrotic syndrome because loss of albumin from the blood decreases blood colloid osmotic pressure. Nephrotic syndrome is associated with several glomerular diseases of unknown cause, as well as with systemic disorders such as diabetes mellitus, systemic lupus erythematosus (SLE), a variety of cancers, and AIDS.

■ Glucosuria

When the blood concentration of glucose is above 200 mg/mL, the renal symporters cannot work fast enough to reabsorb all the glucose that enters the glomerular filtrate. As a result, some glucose remains in the urine, a condition called **glucosuria** (gloo′-kō-SOO-rē-a). The most common cause of glucosuria is diabetes mellitus, in which the blood glucose level may rise far above normal because insulin activity is deficient. Rare genetic mutations in the renal Na^+–glucose symporter greatly reduce its T_m and cause glucosuria. In these cases, glucose appears in the urine even though the blood glucose level is normal. Excessive glucose in the glomerular filtrate inhibits water reabsorption by kidney tubules. This leads to increased urinary output (polyuria), decreased blood volume, and dehydration.

■ Nephroptosis (Floating Kidney)

Nephroptosis (nef′-rōp-Tō-sis; *ptosis* = falling), or **floating kidney,** is an inferior displacement or dropping of the kidney. It occurs when the kidney slips from its normal position because it is not securely held in place by adjacent organs or its covering of fat. Nephroptosis develops most often in very thin people whose adipose capsule or renal fascia is deficient. It is dangerous because the ureter may kink and block urine flow. The resulting backup of urine puts pressure on the kidney, which damages the tissue. Twisting of the ureter also causes pain. Nephroptosis is very common; about one in four people has some degree of weakening of the fibrous bands that hold the kidney in place. It is 10 times more common in females than males. Because it happens during life it is very easy to distinguish from congenital anomalies.

■ Polycystic Kidney Disease

Polycystic kidney disease (PKD) is one of the most common inherited disorders. In PKD, the kidney tubules become riddled with hundreds or thousands of cysts (fluid-filled cavities). In addition, inappropriate apoptosis (programmed cell death) of cells in noncystic tubules leads to progressive impairment of renal function and eventually to end-stage renal failure.

People with PKD also may have cysts and apoptosis in the liver, pancreas, spleen, and gonads; increased risk of cerebral aneurysms; heart valve defects; and diverticuli in the colon. Typically, symptoms are not noticed until adulthood, when patients may have back pain, urinary tract infections, blood in the urine, hypertension, and large abdominal masses. Using drugs to restore normal blood pressure, restricting protein and salt in the diet, and controlling urinary tract infections may slow progression to renal failure.

■ Renal Calculi

The crystals of salts present in urine occasionally precipitate and solidify into insoluble stones called **renal calculi** (*calculi* = pebbles) or **kidney stones.** They commonly contain crystals of calcium oxalate, uric acid, or calcium phosphate. Conditions leading to calculus formation include the ingestion of excessive calcium, low water intake, abnormally alkaline or acidic urine, and overactivity of the parathyroid glands. When a stone lodges in a narrow passage, such as a ureter, the pain can be intense. **Shock-wave lithotripsy** (LITH-ō-trip′-sē; *litho* = stone) is a procedure that uses high-energy shock waves to disintegrate kidney stones and offers an alternative to surgical removal. Once the kidney stone is located using x-rays, a device called a *lithotripter* delivers brief, high-intensity sound waves through a water- or gel-filled cushion placed under the back. Over a period of 30 to 60 minutes, 1000 or more shock waves pulverize the stone, creating fragments that are small enough to wash out in the urine.

■ Renal Failure

Renal failure is a decrease or cessation of glomerular filtration. In **acute renal failure (ARF),** the kidneys abruptly stop working entirely (or almost entirely). The main feature of ARF is the suppression of urine flow, usually characterized either by *oliguria* (daily urine output between 50 mL and 250 mL), or by *anuria* (daily urine output less than 50 mL). Causes include low blood volume (for example, due to hemorrhage), decreased cardiac output, damaged renal tubules, kidney stones, the dyes used to visualize blood vessels in angiograms, nonsteroidal anti-inflammatory drugs, and some antibiotic drugs. It is also common in people who suffer a devastating illness or overwhelming traumatic injury; in such cases it may be related to a more general organ failure known as *multiple organ dysfunction syndrome* (*MODS*).

Renal failure causes a multitude of problems. There is edema due to salt and water retention and metabolic acidosis due to an inability of the kidneys to excrete acidic substances. In the blood, urea builds up due to impaired renal excretion of metabolic waste products and potassium level rises, which can lead to cardiac arrest. Often, there is anemia because the kidneys no longer produce enough erythropoietin for adequate red blood cell production. Because the kidneys are no longer able to convert vitamin D to calcitriol, which is needed for adequate calcium absorption from the small intestine, osteomalacia also may occur.

Chronic renal failure (CRF) refers to a progressive and usually irreversible decline in glomerular filtration rate (GFR). CRF may result from chronic glomerulonephritis, pyelonephritis, polycystic kidney disease, or traumatic loss of kidney tissue. CRF develops in three stages. In the first stage, *diminished renal reserve*, nephrons are destroyed until about 75% of the functioning nephrons are lost. At this stage, a person may have no signs or symptoms because the remaining nephrons enlarge and take over the function of those that have been lost. Once 75% of the nephrons are lost, the person enters the second stage, called *renal insufficiency*, characterized by a decrease in GFR and increased blood levels of nitrogen-containing wastes and creatinine. Also, the kidneys cannot effectively concentrate or dilute the urine. The final stage, called end-stage renal failure, occurs when about 90% of the nephrons have been lost. At this stage, GFR diminishes to 10–15% of normal, oliguria is present, and blood levels of nitrogen-containing wastes and creatinine increase further. People with end-stage renal failure need dialysis therapy and are possible candidates for a kidney transplant operation.

■ Urinary Bladder Cancer

Each year, nearly 12,000 Americans die from **urinary bladder cancer.** It generally strikes people over 50 years of age and is three times more likely to develop in males than females. The disease is typically painless as it develops, but in most cases blood in the urine is a primary sign of the disease. Less often, people experience painful and/or frequent urination.

As long as the disease is identified early and treated promptly, the prognosis is favorable. Fortunately, about 75% of the urinary bladder cancers are confined to the epithelium of the urinary bladder and are easily removed by surgery. The lesions tend to be low-grade, meaning that they have only a small potential for metastasis.

Urinary bladder cancer is frequently the result of a carcinogen. About half of all cases occur in people who smoke or have at some time smoked cigarettes. The cancer also tends to develop in people who are exposed to chemicals called aromatic amines. Workers in the leather, dye, rubber, and aluminum industries, as well as painters, are often exposed to these chemicals.

■ Urinary Tract Infections

The term **urinary tract infection (UTI)** is used to describe either an infection of a part of the urinary system or the presence of large numbers of microbes in urine. UTIs are more common in females due to the shorter length of the urethra. Symptoms include painful or burning urination, urgent and frequent urination, low back pain, and bed-wetting. UTIs include *urethritis* (inflammation of the urethra), *cystitis* (inflammation of the urinary bladder), and *pyelonephritis* (inflammation of the kidneys). If pyelonephritis becomes chronic, scar tissue can form in the kidneys and severely impair their function. Drinking cranberry juice can prevent the attachment of *E. coli* bacteria to the lining of the urinary bladder so that they are more readily flushed away during urination. ■

Additional Clinical Considerations

■ Kidney Transplant

A kidney transplant is the transfer of a kidney from a donor to a recipient whose kidneys no longer function. In the procedure, the donor kidney is placed in the pelvis of the recipient through an abdominal incision. The renal artery and vein of the transplanted kidney are attached to a nearby artery or vein in the pelvis of the recipient and the ureter of the transplanted kidney is then attached to the urinary bladder. During a kidney transplant, the patient receives only one donor kidney, since only one kidney is needed to maintain sufficient renal function. The nonfunctioning diseased kidneys are usually left in place. As with all organ transplants, kidney transplant recipients must be ever vigilant for signs of infection or organ rejection. The transplant recipient will take immunosuppressive drugs for the rest of his or her life to avoid rejection of the "foreign" organ.

■ Loss of Plasma Proteins in Urine Causes Edema

In some kidney diseases, glomerular capillaries are damaged and become so permeable that plasma proteins enter glomerular filtrate. As a result, the filtrate exerts a colloid osmotic pressure that draws water out of the blood. In this situation, the NFP increases, which means more fluid is filtered. At the same time, blood colloid osmotic pressure decreases because plasma proteins are being lost in the urine. Because more fluid filters out of blood capillaries into tissues throughout the body than returns via reabsorption, blood volume decreases and interstitial fluid volume increases. Thus, loss of plasma

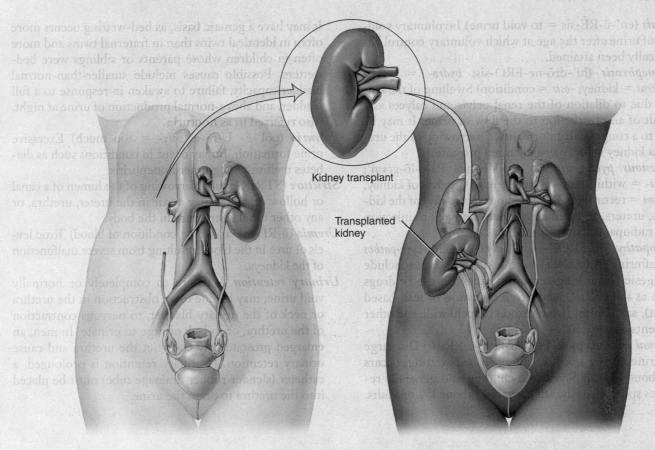

Donor: Functioning kidneys

Recipient: Functioning kidneys

Kidney transplant

Transplanted kidney

proteins in urine causes *edema*, an abnormally high volume of interstitial fluid.

■ Urinary Incontinence

A lack of voluntary control over micturition is called **urinary incontinence.** In infants and children under 2–3 years old, incontinence is normal because neurons to the external urethral sphincter muscle are not completely developed; voiding occurs whenever the urinary bladder is sufficiently distended to stimulate the micturition reflex. Urinary incontinence also occurs in adults. There are four types of urinary incontinence—stress, urge, overflow, and functional. **Stress incontinence** is the most common type of incontinence in young and middle-aged females, and results from weakness of the deep muscles of the pelvic floor. As a result, any physical stress that increases abdominal pressure, such as coughing, sneezing, laughing, exercising, straining, lifting heavy objects, and pregnancy, causes leakage of urine from the urinary bladder. **Urge incontinence** is most common in older people and is characterized by an abrupt and intense urge to urinate followed by an involuntary loss of urine. It may be caused by irritation of the urinary bladder wall by infection or kidney stones, stroke, multiple sclerosis, spinal cord injury, or anxiety. **Overflow incontinence** refers to the involuntary leakage of small amounts of urine caused by some type of blockage or weak contractions of the musculature of the urinary bladder. When urine flow is blocked (for example, from an enlarged prostate or kidney stones) or the urinary bladder muscles can no longer contract, the urinary bladder becomes overfilled and the pressure inside increases until small amounts of urine dribble out. **Functional incontinence** is urine loss resulting from the inability to get to a toilet facility in time as a result of conditions such as stroke, severe arthritis, and Alzheimer disease. Choosing the right treatment option depends on correct diagnosis of the type of incontinence. Treatments include Kegel exercises, urinary bladder training, medication, and possibly even surgery.

Medical Terminology

Azotemia (az-ō-TĒ-mē-a; *azot-* = nitrogen; *-emia* = condition of blood) Presence of urea or other nitrogen-containing substances in the blood.

Cystocele (SIS-tō-sēl; *cysto-* = bladder; *-cele* = hernia or rupture) Hernia of the urinary bladder.

Diabetic kidney disease A disorder caused by diabetes mellitus in which glomeruli are damaged. The result is the leakage of proteins into the urine and a reduction in the ability of the kidney to remove water and waste.

Dysuria (dis-Ū-rē-a; *dys-* = painful; *uria* = urine) Painful urination.

Enuresis (en′-ū-RĒ-sis = to void urine) Involuntary voiding of urine after the age at which voluntary control has typically been attained.

Hydronephrosis (hī′-drō-ne-FRŌ-sis; *hydro-* = water; *nephros* = kidney; *-osis* = condition) Swelling of the kidney due to dilation of the renal pelvis and calyces as a result of an obstruction to the flow of urine. It may be due to a congenital abnormality, a narrowing of the ureter, a kidney stone, or an enlarged prostate.

Intravenous pyelogram (in′-tra-VĒ-nus PĪ-e-lō-gram′; *intra-* = within; *veno-* = vein; *pyelo-* = pelvis of kidney; *-gram* = record) (or **IVP**) Radiograph (x-ray) of the kidneys, ureters, and urinary bladder after venous injection of a radiopaque contrast medium.

Nephropathy (ne-FROP-a-thē; *neph-* = kidney; *-pathos* = suffering) Any disease of the kidneys. Types include analgesic (from long-term and excessive use of drugs such as ibuprofen), lead (from ingestion of lead-based paint), and solvent (from carbon tetrachloride and other solvents).

Nocturnal enuresis (nok-TUR-nal en′-ū-RĒ-sis) Discharge of urine during sleep, resulting in bed-wetting; occurs in about 15% of 5-year-old children and generally resolves spontaneously, afflicting only about 1% of adults.

It may have a genetic basis, as bed-wetting occurs more often in identical twins than in fraternal twins and more often in children whose parents or siblings were bed-wetters. Possible causes include smaller-than-normal bladder capacity, failure to awaken in response to a full bladder, and above-normal production of urine at night. Also referred to as **nocturia.**

Polyuria (pol′-ē- Ū-rē-a; *poly-* = too much) Excessive urine formation. It may occur in conditions such as diabetes mellitus and glomerulonephritis.

Stricture (STRIK-chur) Narrowing of the lumen of a canal or hollow organ, as may occur in the ureter, urethra, or any other tubular structure in the body.

Uremia (ū-RĒ-mē-a; *emia* = condition of blood) Toxic levels of urea in the blood resulting from severe malfunction of the kidneys.

Urinary retention A failure to completely or normally void urine; may be due to an obstruction in the urethra or neck of the urinary bladder, to nervous contraction of the urethra, or to lack of urge to urinate. In men, an enlarged prostate may constrict the urethra and cause urinary retention. If urinary retention is prolonged, a catheter (slender rubber drainage tube) must be placed into the urethra to drain the urine.

Chapter 14

The Reproductive Systems

The **male and female reproductive organs** work together to produce offspring. In addition, the female reproductive organs contribute to sustaining the growth of embryos and fetuses. *Gynecology* is the specialized branch of medicine concerned with the diagnosis and treatment of diseases of the female reproductive system. *Urologists* diagnose and treat diseases and disorders of the male reproductive system. The branch of medicine that deals with male disorders, especially infertility and sexual dysfunction is called *andrology*.

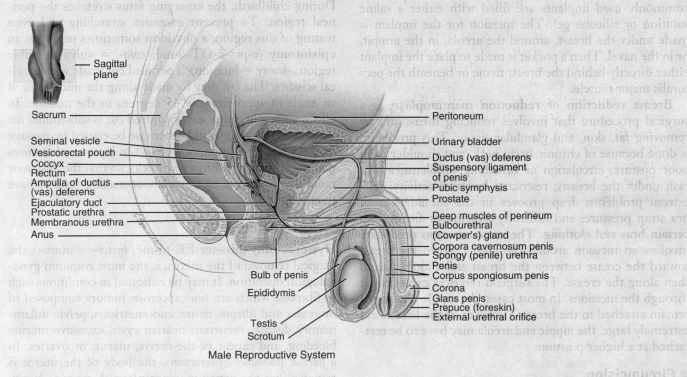

Sagittal plane

- Sacrum
- Seminal vesicle
- Vesicorectal pouch
- Coccyx
- Rectum
- Ampulla of ductus (vas) deferens
- Ejaculatory duct
- Prostatic urethra
- Membranous urethra
- Anus
- Bulb of penis
- Epididymis
- Testis
- Scrotum

- Peritoneum
- Urinary bladder
- Ductus (vas) deferens
- Suspensory ligament of penis
- Pubic symphysis
- Prostate
- Deep muscles of perineum
- Bulbourethral (Cowper's) gland
- Corpora cavernosum penis
- Spongy (penile) urethra
- Penis
- Corpus spongiosum penis
- Corona
- Glans penis
- Prepuce (foreskin)
- External urethral orifice

Male Reproductive System

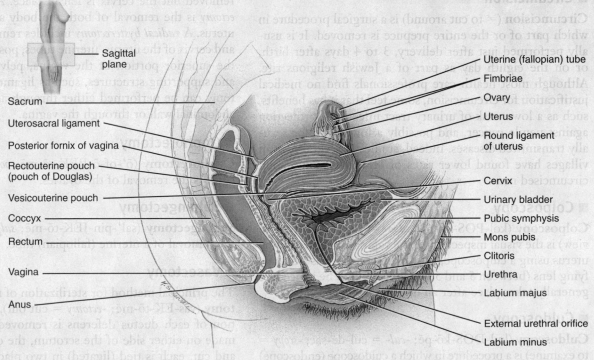

Sagittal plane

- Sacrum
- Uterosacral ligament
- Posterior fornix of vagina
- Rectouterine pouch (pouch of Douglas)
- Vesicouterine pouch
- Coccyx
- Rectum
- Vagina
- Anus

- Uterine (fallopian) tube
- Fimbriae
- Ovary
- Uterus
- Round ligament of uterus
- Cervix
- Urinary bladder
- Pubic symphysis
- Mons pubis
- Clitoris
- Urethra
- Labium majus
- External urethral orifice
- Labium minus

Female Reproductive System

Tests, Procedures, and Techniques

■ Breast Augmentation and Reduction

Breast augmentation (enlargement), technically called **augmentation mammoplasty,** is a surgical procedure to increase breast size and shape. It may be done to enhance breast size for females who feel that their breasts are too small, to restore breast volume due to weight loss or following pregnancy, to improve the shape of breasts that are sagging, and to improve breast appearance following surgery, trauma, or congenital abnormalities. The most commonly used implants are filled with either a saline solution or silicone gel. The incision for the implant is made under the breast, around the areola, in the armpit, or in the navel. Then a pocket is made to place the implant either directly behind the breast tissue or beneath the pectoralis major muscle.

Breast reduction or **reduction mammoplasty** is a surgical procedure that involves reducing breast size by removing fat, skin, and glandular tissue. This procedure is done because of chronic back, neck, and shoulder pain; poor posture; circulation or breathing problems; a skin rash under the breasts; restricted levels of activity; self-esteem problems; deep grooves in the shoulders from bra strap pressure; and difficulty wearing or fitting into certain bras and clothing. The most common procedure involves an incision around the areola, down the breast toward the crease between the breast and abdomen, and then along the crease. The surgeon removes excess tissue through the incisions. In most cases, the nipple and areola remain attached to the breast. However, if the breasts are extremely large, the nipple and areola may have to be reattached at a higher position.

■ Circumcision

Circumcision (= to cut around) is a surgical procedure in which part of or the entire prepuce is removed. It is usually performed just after delivery, 3 to 4 days after birth, or on the eighth day as part of a Jewish religious rite. Although most health-care professionals find no medical justification for circumcision, some feel that it has benefits, such as a lower risk of urinary tract infections, protection against penile cancer, and possibly a lower risk for sexually transmitted diseases. Indeed, studies in several African villages have found lower rates of HIV infection among circumcised men.

■ Colposcopy

Colposcopy (kol-POS-kō-pē; *colpo-* = vagina; *-scopy* = to view) is the visual inspection of the vagina and cervix of the uterus using a culposcope, an instrument that has a magnifying lens (between 5 and 50 ×) and a light. The procedure generally takes place after an unusual Pap smear.

■ Culdoscopy

Culdoscopy (kul-DOS-kō-pē; *-cul-* = cul-de-sac; *-scopy* = to examine) is a procedure in which a culdoscope (endoscope) is inserted through the posterior wall of the vagina to view the rectouterine pouch in the pelvic cavity.

■ Endocervical curettage

Endocervical curettage(kū-re-TAHZH; *curette* = scraper) is a procedure in which the cervix is dilated and the endometrium of the uterus is scraped with a spoon-shaped instrument called a curette; commonly called a *D and C* (dilation and curettage).

■ Episiotomy

During childbirth, the emerging fetus stretches the perineal region. To prevent excessive stretching and even tearing of this region, a physician sometimes performs an **episiotomy** (e-piz-ē-OT-ō-mē; *episi-* = vulva or pubic region; *-otomy* = incision), a perineal cut made with surgical scissors. The cut may be made along the midline or at an angle of approximately 45 degrees to the midline. In effect, a straight, more easily sutured cut is substituted for the jagged tear that would otherwise be caused by passage of the fetus. The incision is closed in layers with sutures that are absorbed within a few weeks, so that the busy new mom does not have to worry about making time to have them removed.

■ Hysterectomy

Hysterectomy (hiss-ter-EK-tō-mē; *hyster-* = uterus), the surgical removal of the uterus, is the most common gynecological operation. It may be indicated in conditions such as fibroids, which are noncancerous tumors composed of muscular and fibrous tissue, endometriosis, pelvic inflammatory disease, recurrent ovarian cysts, excessive uterine bleeding, and cancer of the cervix, uterus, or ovaries. In a *partial (subtotal) hysterectomy*, the body of the uterus is removed but the cervix is left in place. A *complete hysterectomy* is the removal of both the body and cervix of the uterus. A *radical hysterectomy* includes removal of the body and cervix of the uterus, uterine tubes, possibly the ovaries, the superior portion of the vagina, pelvic lymph nodes, and supporting structures, such as ligaments. A hysterectomy can be performed either through an incision in the abdominal wall or through the vagina.

■ Oophorectomy

Oophorectomy (ō′-of-ō-REK-tō-mē; *oophor-* = bearing eggs) is the removal of the ovaries.

■ Salpingectomy

Salpingectomy (sal′-pin-JEK-tō-mē; *salpingo* = tube) is the removal of a uterine (fallopian) tube.

■ Vasectomy

The principal method for sterilization of males is a **vasectomy** (vas-EK-tō-mē; *-ectomy* = cut out), in which a portion of each ductus deferens is removed. An incision is made on either side of the scrotum, the ducts are located and cut, each is tied (ligated) in two places with stitches, and the portion between the ties is removed. Although sperm production continues in the testes, sperm can no longer reach the exterior. The sperm degenerate and are destroyed by phagocytosis. Because the blood vessels are not cut, testosterone levels in the blood remain normal, so vasectomy has no effect on sexual desire, performance, and ejaculation. If done correctly, it is close to 100% effective. The procedure can be reversed, but the chance of regaining fertility is only 30–40%.

Disorders Affecting the Reproductive Systems

■ Breast Cancer

One in eight women in the United States faces the prospect of **breast cancer.** After lung cancer, it is the second-leading cause of death from cancer in U.S. women. Breast cancer can occur in males but is rare. In females, breast cancer is seldom seen before age 30; its incidence rises rapidly after menopause. An estimated 5% of the 180,000 cases diagnosed each year in the United States, particularly those that arise in younger women, stem from inherited genetic mutations (changes in the DNA). Researchers have now identified two genes that increase susceptibility to breast cancer: *BRCA*1 (*br*east *c*ancer 1) and *BRCA*2. Mutation of *BRCA*1 also confers a high risk for ovarian cancer. In addition, mutations of the *p53* gene increase the risk of breast cancer in both males and females, and mutations of the androgen receptor gene are associated with the occurrence of breast cancer in some males. Because breast cancer generally is not painful until it becomes quite advanced, any lump, no matter how small, should be reported to a physician at once. Early detection—by breast self-examination and mammograms—is the best way to increase the chance of survival.

The most effective technique for detecting tumors less than 1 cm (0.4 in.) in diameter is **mammography** (mam-OG-ra-fē; *-graphy* = to record), a type of radiography using very sensitive x-ray film. The image of the breast, called a **mammogram**, is best obtained by compressing the breasts, one at a time, using flat plates. A supplementary procedure for evaluating breast abnormalities is **ultrasound.** Although ultrasound cannot detect tumors smaller than 1 cm in diameter (which mammography can detect), it can be used to determine whether a lump is a benign, fluid-filled cyst or a solid (and therefore possibly malignant) tumor.

Among the factors that increase the risk of developing breast cancer are (1) a family history of breast cancer, especially in a mother or sister; (2) nulliparity (never having borne a child) or having a first child after age 35; (3) previous cancer in one breast; (4) exposure to ionizing radiation, such as x-rays; (5) excessive alcohol intake; and (6) cigarette smoking.

The American Cancer Society recommends the following steps to help diagnose breast cancer as early as possible:

- All women over 20 should develop the habit of monthly breast self-examination.
- A physician should examine the breasts every 3 years when a woman is between the ages of 20 and 40, and every year after age 40.
- A mammogram should be taken in women between the ages of 35 and 39, to be used later for comparison (baseline mammogram).
- Women with no symptoms should have a mammogram every year after age 40.
- Women of any age with a history of breast cancer, a strong family history of the disease, or other risk factors should consult a physician to determine a schedule for mammography.

Treatment for breast cancer may involve hormone therapy, chemotherapy, radiation therapy, **lumpectomy** (removal of the tumor and the immediate surrounding tissue), a modified or radical mastectomy, or a combination of these approaches. A **radical mastectomy** (*mast-* = breast) involves removal of the affected breast along with the underlying pectoral muscles and the axillary lymph nodes. (Lymph nodes are removed because metastasis of cancerous cells usually occurs through lymphatic or blood vessels.) Radiation treatment and chemotherapy may follow the surgery to ensure the destruction of any stray cancer cells. Several types of chemotherapeutic drugs are used to decrease the risk of relapse or disease progression. *Nolvadex*®(*tamoxifen*) is an estrogen antagonist that binds to and blocks estrogen receptors, thus decreasing the stimulating effect of estrogen on breast cancer cells. Tamoxifen has been used for 20 years and greatly reduces the risk of cancer recurrence. *Herceptin*®, a monoclonal antibody drug, targets an antigen on the surface of breast cancer cells. It is effective in causing regression of tumors and retarding progression of the disease. The early data from clinical trials of two new drugs, *Femara*® and *Amimidex*®, show relapse rates that are lower than those for tamoxifen. These drugs are inhibitors of aromatase, the enzyme needed for the final step in synthesis of estrogens. Finally, two drugs—tamoxifen and *Evista*®(*raloxifene*)—are being marketed for breast cancer *prevention*. Interestingly, raloxifene blocks estrogen receptors in the breasts and uterus but activates estrogen receptors in bone, providing effective treatment for osteoporosis while possibly decreasing the risk of breast or endometrial (uterine) cancer.

■ Cervical Cancer

Cervical cancer, carcinoma of the cervix of the uterus, starts with **cervical dysplasia** (dis-PLĀ-sē-a), a change in the shape, growth, and number of cervical cells. The cells may either return to normal or progress to cancer. In most cases, cervical cancer may be detected in its earliest stages by a Pap test. Some evidence links cervical cancer to

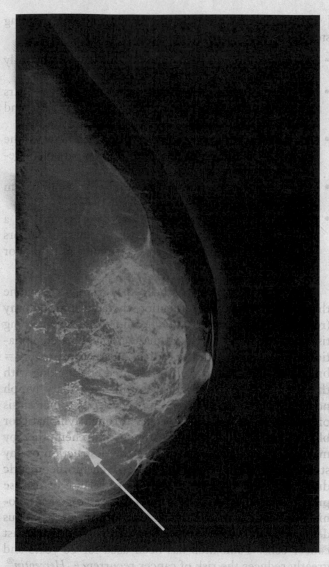

Mammogram showing breast cancer (white spot)

■ Endometriosis

Endometriosis (en-dō-mē-trē-Ō-sis; *endo-* = within; *metri-* = uterus; *osis* = condition) is characterized by the growth of endometrial tissue outside the uterus. The tissue enters the pelvic cavity via the open uterine tubes and may be found in any of several sites—on the ovaries, the rectouterine pouch, the outer surface of the uterus, the sigmoid colon, pelvic and abdominal lymph nodes, the cervix, the abdominal wall, the kidneys, and the urinary bladder. Endometrial tissue responds to hormonal fluctuations, whether it is inside or outside the uterus. With each reproductive cycle, the tissue proliferates and then breaks down and bleeds. When this occurs outside the uterus, it can cause inflammation, pain, scarring, and infertility. Symptoms include premenstrual pain or unusually severe menstrual pain.

■ Female Athlete Triad: Disordered Eating, Amenorrhea, and Premature Osteoporosis

The female reproductive cycle can be disrupted by many factors, including weight loss, low body weight, disordered eating, and vigorous physical activity. The observation that three conditions—disordered eating, amenorrhea, and osteoporosis—often occur together in female athletes led researchers to coin the term **female athlete triad.**

Many athletes experience intense pressure from coaches, parents, peers, and themselves to lose weight to improve performance. Hence, they may develop disordered eating behaviors and engage in other harmful weight-loss practices in a struggle to maintain a very low body weight. **Amenorrhea** (ā-men′-ō-RĒ-a; *a-* = without; *men-* = month; *-rrhea* = a flow) is the absence of menstruation. The most common causes of amenorrhea are pregnancy and menopause. In female athletes, amenorrhea results from reduced secretion of gonadotropin-releasing hormone, which decreases the release of LH and FSH. As a result, ovarian follicles fail to develop, ovulation does not occur, synthesis of estrogens and progesterone wanes, and monthly menstrual bleeding ceases. Most cases of the female athlete triad occur in young women with very low amounts of body fat. Low levels of the hormone leptin, secreted by adipose cells, may be a contributing factor.

Because estrogens help bones retain calcium and other minerals, chronically low levels of estrogens are associated with loss of bone mineral density. The female athlete triad causes "old bones in young women." In one study, amenorrheic runners in their twenties had low bone mineral densities, similar to those of postmenopausal women 50 to 70 years old! Short periods of amenorrhea in young athletes may cause no lasting harm. However, long-term cessation of the reproductive cycle may be accompanied by a loss of bone mass, and adolescent athletes may fail to achieve an adequate bone mass; both of these situations can lead to premature osteoporosis and irreversible bone damage.

the virus that causes genital warts, human papillomavirus (HPV). Increased risk is associated with having a large number of sexual partners, having first intercourse at a young age, and smoking cigarettes.

■ Cryptorchidism

The condition in which the testes do not descend into the scrotum is called **cryptorchidism** (krip-TOR-ki-dizm; *crypt-* = hidden; *orchid* = testis); it occurs in about 3% of full-term infants and about 30% of premature infants. Untreated bilateral cryptorchidism results in sterility because the cells involved in the initial stages of spermatogenesis are destroyed by the higher temperature of the pelvic cavity. The chance of testicular cancer is 30–50 times greater in cryptorchid testes. The testes of about 80% of boys with cryptorchidism will descend spontaneously during the first year of life. When the testes remain undescended, the condition can be corrected surgically, ideally before 18 months of age.

Fibrocystic Disease of the Breasts

The breasts of females are highly susceptible to cysts and tumors. In **fibrocystic disease,** the most common cause of breast lumps in females, one or more cysts (fluid-filled sacs) and thickenings of alveoli develop. The condition, which occurs mainly in females between the ages of 30 and 50, is probably due to a relative excess of estrogens or a deficiency of progesterone in the postovulatory (luteal) phase of the reproductive cycle (discussed shortly). Fibrocystic disease usually causes one or both breasts to become lumpy, swollen, and tender a week or so before menstruation begins.

Ovarian Cancer

Even though **ovarian cancer** is the sixth most common form of cancer in females, it is the leading cause of death from all gynecological malignancies (excluding breast cancer) because it is difficult to detect before it metastasizes (spreads) beyond the ovaries. Risk factors associated with ovarian cancer include age (usually over age 50); race (whites are at highest risk); family history of ovarian cancer; more than 40 years of active ovulation; nulliparity or first pregnancy after age 30; a high-fat, low-fiber, vitamin A–deficient diet; and prolonged exposure to asbestos or talc. Early ovarian cancer has no symptoms or only mild ones associated with other common problems, such as abdominal discomfort, heartburn, nausea, loss of appetite, bloating, and flatulence. Later-stage signs and symptoms include an enlarged abdomen, abdominal and/or pelvic pain, persistent gastrointestinal disturbances, urinary complications, menstrual irregularities, and heavy menstrual bleeding.

Pelvic inflammatory disease

Pelvic inflammatory disease (PID*)* A collective term for any extensive bacterial infection of the pelvic organs, especially the uterus, uterine tubes, or ovaries, which is characterized by pelvic soreness, lower back pain, abdominal pain, and urethritis. Often the early symptoms of PID occur just after menstruation. As infection spreads, fever may develop, along with painful abscesses of the reproductive organs.

Premenstrual Syndrome and Premenstrual Dysphoric Disorder

Premenstrual syndrome (PMS) is a cyclical disorder of severe physical and emotional distress. It appears during the postovulatory (luteal) phase of the female reproductive cycle and dramatically disappears when menstruation begins. The signs and symptoms are highly variable from one woman to another. They may include edema, weight gain, breast swelling and tenderness, abdominal distension, backache, joint pain, constipation, skin eruptions, fatigue and lethargy, greater need for sleep, depression or anxiety, irritability, mood swings, headache, poor coordination and clumsiness, and cravings for sweet or salty foods. The cause of PMS is unknown.

For some women, getting regular exercise; avoiding caffeine, salt, and alcohol; and eating a diet that is high in complex carbohydrates and lean proteins can bring considerable relief.

Premenstrual dysphoric disorder (PMDD) is a more severe syndrome in which PMS-like signs and symptoms do not resolve after the onset of menstruation. Clinical research studies have found that suppression of the reproductive cycle by a drug that interferes with GnRH (leuprolide) decreases symptoms significantly. Because symptoms reappear when estradiol or progesterone is given together with leuprolide, researchers propose that PMDD is caused by abnormal responses to normal levels of these ovarian hormones. *SSRIs* (selective serotonin receptor inhibitors) have shown promise in treating both PMS and PMDD.

Prostate Disorders

Because the prostate surrounds part of the urethra, any infection, enlargement, or tumor can obstruct the flow of urine. Acute and chronic infections of the prostate are common in postpubescent males, often in association with inflammation of the urethra. Symptoms may include fever, chills, urinary frequency, frequent urination at night, difficulty in urinating, burning or painful urination, low back pain, joint and muscle pain, blood in the urine, or painful ejaculation. However, often there are no symptoms. Antibiotics are used to treat most cases that result from a bacterial infection. In **acute prostatitis,** the prostate becomes swollen and tender. **Chronic prostatitis** is one of the most common chronic infections in men of the middle and later years. On examination, the prostate feels enlarged, soft, and very tender, and its surface outline is irregular.

Prostate cancer is the leading cause of death from cancer in men in the United States, having surpassed lung cancer in 1991. Each year it is diagnosed in almost 200,000 U.S. men and causes nearly 40,000 deaths. The amount of PSA (prostate-specific antigen), which is produced only by prostate epithelial cells, increases with enlargement of the prostate and may indicate infection, benign enlargement, or prostate cancer. A blood test can measure the level of PSA in the blood. Males over the age of 40 should have an annual examination of the prostate gland. In a **digital rectal exam,** a physician palpates the gland through the rectum with the fingers (digits). Many physicians also recommend an annual PSA test for males over age 50. Treatment for prostate cancer may involve surgery, cryotherapy, radiation, hormonal therapy, and chemotherapy. Because many prostate cancers grow very slowly, some urologists recommend "watchful waiting" before treating small tumors in men over age 70.

Sexually Transmitted Diseases

A **sexually transmitted disease (STD)** is one that is spread by sexual contact. In most developed countries of the world, such as those of Western Europe, Japan,

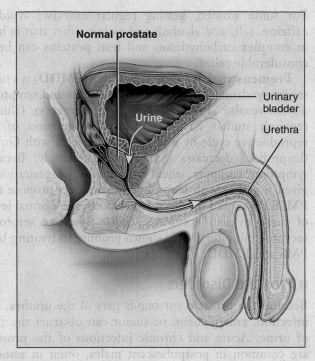

Normal prostate

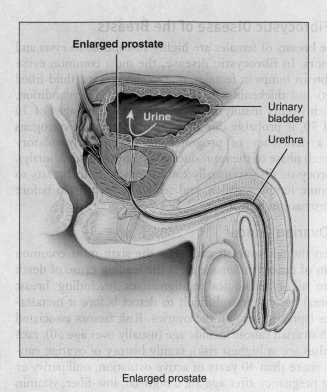

Enlarged prostate

Australia, and New Zealand, the incidence of STDs has declined markedly during the past 25 years. In the United States, by contrast, STDs have been rising to near-epidemic proportions; they currently affect more than 65 million people. AIDS and hepatitis B, which are sexually transmitted diseases that also may be contracted in other ways, are discussed in Chapters 22 and 24, respectively.

Chlamydia

Chlamydia (kla-MID-ē-a) is a sexually transmitted disease caused by the bacterium *Chlamydia trachomatis* (*chlamy-* = cloak). This unusual bacterium cannot reproduce outside body cells; it "cloaks" itself inside cells, where it divides. At present, chlamydia is the most prevalent sexually transmitted disease in the United States. In most cases, the initial infection is asymptomatic and thus difficult to recognize clinically. In males, urethritis is the principal result, causing a clear discharge, burning on urination, frequent urination, and painful urination. Without treatment, the epididymides may also become inflamed, leading to sterility. In 70% of females with chlamydia, symptoms are absent, but chlamydia is the leading cause of pelvic inflammatory disease. The uterine tubes may also become inflamed, which increases the risk of ectopic pregnancy (implantation of a fertilized ovum outside the uterus) and infertility due to the formation of scar tissue in the tubes.

Gonorrhea

Gonorrhea (gon-ō-RĒ-a) or **"the clap"** is caused by the bacterium *Neisseria gonorrhoeae*. In the United States, 1–2 million new cases of gonorrhea appear each year, most

among individuals aged 15–29 years. Discharges from infected mucous membranes are the source of transmission of the bacteria either during sexual contact or during the passage of a newborn through the birth canal. The infection site can be in the mouth and throat after oral–genital contact, in the vagina and penis after genital intercourse, or in the rectum after recto–genital contact.

Males usually experience urethritis with profuse pus drainage and painful urination. The prostate and epididymis may also become infected. In females, infection typically occurs in the vagina, often with a discharge of pus. Both infected males and females may harbor the disease without any symptoms, however, until it has progressed to a more advanced stage; about 5–10% of males and 50% of females are asymptomatic. In females, the infection and consequent inflammation can proceed from the vagina into the uterus, uterine tubes, and pelvic cavity. An estimated 50,000 to 80,000 women in the United States are made infertile by gonorrhea every year as a result of scar tissue formation that closes the uterine tubes. If bacteria in the birth canal are transmitted to the eyes of a newborn, blindness can result. Administration of a 1% silver nitrate solution in the infant's eyes prevents infection.

Syphilis

Syphilis, caused by the bacterium *Treponema pallidum*, is transmitted through sexual contact or exchange of blood, or through the placenta to a fetus. The disease progresses through several stages. During the *primary stage*, the chief sign is a painless open sore, called a **chancre** (SHANG-ker), at the point of contact. The chancre heals within

1 to 5 weeks. From 6 to 24 weeks later, signs and symptoms such as a skin rash, fever, and aches in the joints and muscles usher in the *secondary stage*, which is systemic—the infection spreads to all major body systems. When signs of organ degeneration appear, the disease is said to be in the *tertiary stage*. If the nervous system is involved, the tertiary stage is called **neurosyphilis.** As motor areas become damaged extensively, victims may be unable to control urine and bowel movements. Eventually they may become bedridden and unable even to feed themselves. In addition, damage to the cerebral cortex produces memory loss and personality changes that range from irritability to hallucinations.

■ Genital Herpes

Genital herpes is an incurable STD. Type II herpes simplex virus (HSV-2) causes genital infections, producing painful blisters on the prepuce, glans penis, and penile shaft in males and on the vulva or sometimes high up in the vagina in females. The blisters disappear and reappear in most patients, but the virus itself remains in the body. A related virus, type I herpes simplex virus (HSV-1), causes cold sores on the mouth and lips. Infected individuals typically experience recurrences of symptoms several times a year.

■ Genital Warts

Warts are an infectious disease caused by viruses. *Human papillomavirus (HPV)* causes **genital warts,** which is commonly transmitted sexually. Nearly one million people a year develop genital warts in the United States. Patients with a history of genital warts may be at increased risk for cancers of the cervix, vagina, anus, vulva, and penis. There is no cure for genital warts.

■ Testicular Cancer

Testicular cancer is the most common cancer in males between the ages of 20 and 35. More than 95% of testicular cancers arise from spermatogenic cells within the seminiferous tubules. An early sign of testicular cancer is a mass in the testis, often associated with a sensation of testicular heaviness or a dull ache in the lower abdomen; pain usually does not occur. To increase the chance for early detection of a testicular cancer, all males should perform regular self-examinations of the testes. The examination should be done starting in the teen years and once each month thereafter. After a warm bath or shower (when the scrotal skin is loose and relaxed) each testicle should be examined as follows. The testicle is grasped and gently rolled between the index finger and thumb, feeling for lumps, swellings, hardness, or other changes. If a lump or other change is detected, a physician should be consulted as soon as possible.

■ Vulvovaginal Candidiasis

Candida albicans is a yeastlike fungus that commonly grows on mucous membranes of the gastrointestinal and genitourinary tracts. The organism is responsible for **vulvovaginal candidiasis** (vul-vō-VAJ-i-nal can-di-DĪ-a-sis), the most common form of **vaginitis** (vaj-i-NĪ-tis), inflammation of the vagina. Candidiasis is characterized by severe itching; a thick, yellow, cheesy discharge; a yeasty odor; and pain. The disorder, experienced at least once by about 75% of females, is usually a result of proliferation of the fungus following antibiotic therapy for another condition. Predisposing conditions include the use of oral contraceptives or cortisone-like medications, pregnancy, and diabetes.

Additional Clinical Considerations

■ Erectile Dysfunction

Erectile dysfunction (ED), previously termed *impotence,* is the consistent inability of an adult male to ejaculate or to attain or hold an erection long enough for sexual intercourse. Many cases of impotence are caused by insufficient release of nitric oxide (NO), which relaxes the smooth muscle of the penile arterioles and erectile tissue. The drug *Viagra*® (sildenafil) enhances smooth muscle relaxation by nitric oxide in the penis. Other causes of erectile dysfunction include diabetes mellitus, physical abnormalities of the penis, systemic disorders such as syphilis, vascular disturbances (arterial or venous obstructions), neurological disorders, surgery, testosterone deficiency, and drugs (alcohol, antidepressants, antihistamines, antihypertensives, narcotics, nicotine, and tranquilizers). Psychological factors such as anxiety or depression, fear of causing pregnancy, fear of sexually transmitted diseases, religious inhibitions, and emotional immaturity may also cause ED.

■ Ovarian Cysts

An **ovarian cyst** is a fluid-filled sac in or on an ovary. Such cysts are relatively common, are usually noncancerous, and frequently disappear on their own. Cancerous cysts are more likely to occur in women over 40. Ovarian cysts may cause pain, pressure, a dull ache, or fullness in the abdomen; pain during sexual intercourse; delayed, painful, or irregular menstrual periods; abrupt onset of sharp pain in the lower abdomen; and/or vaginal bleeding. Most ovarian cysts require no treatment, but larger ones (more than 5 cm or 2 in.) may be removed surgically.

■ Premature Ejaculation

A **premature ejaculation** is ejaculation that occurs too early, for example, during foreplay or upon or shortly after penetration. It is usually caused by anxiety, other psychological causes, or an unusually sensitive foreskin or glans penis. For most males, premature ejaculation can be overcome by various techniques (such as squeezing the penis between the glans penis and shaft as ejaculation approaches), behavioral therapy, or medication.

■ Uterine Prolapse

A condition called **uterine prolapse** (*prolapse* = falling down or downward displacement) may result from weakening of supporting ligaments and pelvic musculature associated with age or disease, traumatic vaginal delivery, chronic straining from coughing or difficult bowel movements, or pelvic tumors. The prolapse may be characterized as *first degree* (*mild*), in which the cervix remains within the vagina; *second degree* (*marked*), in which the cervix protrudes through the vagina to the exterior; and *third degree* (*complete*), in which the entire uterus is outside the vagina. Depending on the degree of prolapse, treatment may involve pelvic exercises, dieting if a patient is overweight, a stool softener to minimize straining during defecation, pessary therapy (placement of a rubber device around the uterine cervix that helps prop up the uterus), or surgery.

Medical Terminology

Castration (kas-TRĀ-shun = to prune) Removal, inactivation, or destruction of the gonads; commonly used in reference to removal of the testes only.

Dysmenorrhea (dis′-men-or-Ē-a; *dys-* = difficult or painful) Pain associated with menstruation; the term is usually reserved to describe menstrual symptoms that are severe enough to prevent a woman from functioning normally for one or more days each month. Some cases are caused by uterine tumors, ovarian cysts, pelvic inflammatory disease, or intrauterine devices.

Dyspareunia (dis-pa-ROO-nē-a; *dys-* = difficult; *para* = beside; *-enue* = bed) Pain during sexual intercourse. It may occur in the genital area or in the pelvic cavity, and may be due to inadequate lubrication, inflammation, infection, an improperly fitting diaphragm or cervical cap, endometriosis, pelvic inflammatory disease, pelvic tumors, or weakened uterine ligaments.

Fibroids (FĪ-broyds; *fibro-* = fiber; *-eidos* = resemblance) Noncancerous tumors in the myometrium of the uterus composed of muscular and fibrous tissue. Their growth appears to be related to high levels of estrogens. They do not occur before puberty and usually stop growing after menopause. Symptoms include abnormal menstrual bleeding and pain or pressure in the pelvic area.

Hermaphroditism (her-MAF-rō-dīt-izm) The presence of both ovarian and testicular tissue in one individual.

Hypospadias (hī′-pō-SPĀ-dē-as; *hypo-* = below) A common congenital abnormality in which the urethral opening is displaced. In males, the displaced opening may be on the underside of the penis, at the penoscrotal junction, between the scrotal folds, or in the perineum; in females, the urethra opens into the vagina. The problem can be corrected surgically.

Leukorrhea (loo′-kō-RĒ-a; *leuko-* = white) A whitish (nonbloody) vaginal discharge containing mucus and pus cells that may occur at any age and affects most women at some time.

Menorrhagia (men-ō-RA-jē-a; *meno-* = menstruation; -*rhage* = to burst forth) Excessively prolonged or profuse menstrual period. May be due to a disturbance in hormonal regulation of the menstrual cycle, pelvic infection, medications (anticoagulants), fibroids (noncancerous uterine tumors composed of muscle and fibrous tissue), endometriosis, or intrauterine devices.

Orchitis (or-KĪ-tis; *orchi-* = testes; *-itis* = inflammation) Inflammation of the testes, for example, as a result of the mumps virus or a bacterial infection.

Smegma (SMEG-ma) The secretion, consisting principally of desquamated epithelial cells, found chiefly around the external genitals and especially under the foreskin of the male.

Photo Credits

Chapter 1 Page 2 (top left): Scientifica/Visuals Unlimited. Page 2 (bottom left): @Mark Saior/Phototake. Page 2 (bottom right): Jose Luis Pelaez/Getty Images, Inc. Page 2 (bottom right): Matrix/Medical Photo Researchers, Inc. Page 3 (top left): Biophoto Associates/Photo Researchers. Page 3 (top center): Breast Cancer Unit, Kings College Hospital, London/Photo Researchers, Inc. Page 3 (top right): Zephyr/Photo Researchers, Inc. Page 3 (bottom left): Cardio-Thoracic Centre, Freeman Hospital, Newcastle-Upon-Tyne/Photo Researchers, Inc. Page 3 (bottom center): CNRI/Science Photo Library/Photo Researchers, Inc. Page 3 (bottom right): Science Photo Library/Photo Researchers, Inc. Page 4 (top left): Scott Camazine/Photo Researchers, Inc. Page 4 (top right): Simon Fraser/Photo Researchers, Inc. Page 4 (bottom right): Courtesy Andrew Joseph Torres and Dennis Soler. Page 5 (top right): Howard Sochurek/Medical Images, Inc. Page 5 (center left): SIU/Visuals Unlimited. Page 5 (center right): Dept. of Nuclear Medicine, Charing Cross Hospital/Photo Researchers, Inc. Page 5 (bottom right): @Camal/Phototake.

Chapter 2 Page 7 (center left): Russell Goodenbache/Photo Researchers, Inc. Page 12 (bottom right): U.S. Department of Energy Genomic Programs.

Chapter 3 Page 17 (top left): Sheila Terry/Science Photo Library/Photo Researchers, Inc. Page 17 (top center): St. Stephen's Hospital, London/Science Photo Library/Photo Researchers, Inc. Page 17 (top right): St. Stephen's Hospital/Science Photo Library/Photo Researchers, Inc. Page 18: Dr. P. Marazzi/Science Photo Library/Photo Researchers, Inc. Page 19 (top left): Alain Dex/Photo Researchers, Inc. Page 19 (top right): Biophoto Associates/Photo Researchers, Inc.

Chapter 4 Page 23: CNRI/Photo Researchers, Inc. Page 24 (top left): Princess Margaret Rose Orthopaedic Hospital/Photo Researchers, Inc. Page 24 (top center): Dr. G. Marazzi/Photo Researchers, Inc. Page 24 (top right): Custom Medical Stock Photo, Inc. Page 25 (top right): P. Morton/Photo Researchers, Inc. Page 25 (top right): P. Morton/Photo Researchers, Inc. Page 26: Centers for Disease Control. Page 27 (top): @ISM/Phototake/Phototake. Page 27 (bottom): Neil Borden/Photo Researchers, Inc.

Chapter 5 Page 31 (top center): SIU/Visuals Unlimited. Page 31 (bottom right): ISM/Phototake. Page 32 (center right): Charles McRae/Visuals Unlimited. Page 33: CNRI/Photo Researchers, Inc. Page 34: @Biophoto/CMP Images/Phototake.

Chapter 7 Page 45: Image Source/Getty Images. Page 49: Steve Allen/Photo Researchers, Inc.

Chapter 8 Page 62: Biophoto Associates/Photo Researchers, Inc. Page 63: CNRI/Phototake. Page 64 (top left): Pulse Picture Library/CMP Images/Phototake. Page 64 (top right): Ian Boddy/Photo Researchers, Inc. Page 65 (top left): from New England Journal of Medicine, February 18, 1999, vol. 340, No. 7. Page 234: Photo provided courtesy of Robert Capel, Department of Internal Medicine, University of Texas, M.D. Anderson Cancer Center, Houston, Texas. Page 65 (bottom left): Laser Bergman/The Bergman Collection. Page 66 (top left): Laser Bergman/The Bergman Collection. Page 66 (top right): Laser Bergman/The Bergman Collection.

Chapter 9 Page 67 (top right): Jason Kryger/Photo Researchers, Inc. Page 67: Mark Nielsen. Page 68: Scott Camazine/Phototake. Page 72 (bottom left): Carolina Biological Supply/Phototake. Page 72 (bottom right): W. Ober/Visuals Unlimited. Page 73: ISM/Phototake. Page 80 (center left): Stanley Flegler/Visuals Unlimited. Page 80 (bottom right): Scott Camazine/Phototake.

Chapter 10 Page 89: P. Marazzi/Photo Researchers, Inc.

Chapter 11 Page 93: BSIP/Phototake. Page 94: Terry Williams/The Image Bank/Getty Images. Page 95: Jocelyn Berger/Photo Researchers, Inc.

Chapter 12 Page 100: David M. Martin/Photo Researchers, Inc.

Chapter 13 Page 106: BSIP/Phototake.

Chapter 14 Page 114: ISM/Phototake.